THE UNIVERSITY OF MICHIGAN

CENTER FOR SOUTH AND SOUTHEAST ASIAN STUDIES

MICHIGAN PAPERS ON SOUTH AND SOUTHEAST ASIA

Ann Arbor, Michigan
USA

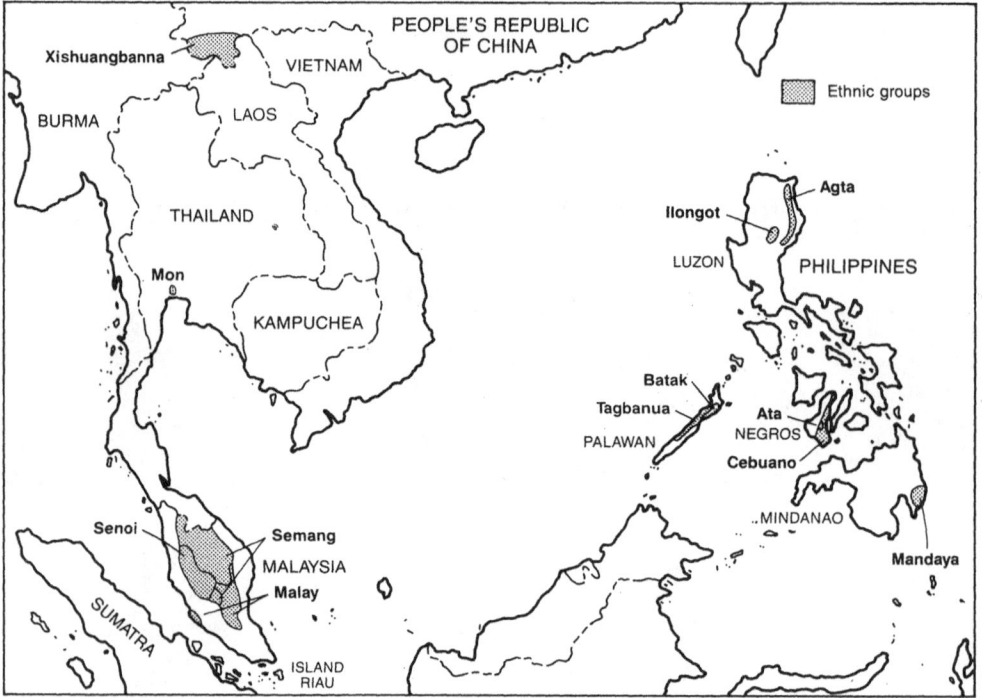

Location of groups discussed in this volume (drawn by K. Gillogly)

ETHNIC DIVERSITY AND THE CONTROL OF NATURAL RESOURCES IN SOUTHEAST ASIA

edited by A. Terry Rambo, Kathleen Gillogly, and Karl L. Hutterer

Published in cooperation with the East-West Center, Environment and Policy Institute, Honolulu, Hawaii

MICHIGAN PAPERS ON SOUTH AND SOUTHEAST ASIA
CENTER FOR SOUTH AND SOUTHEAST ASIAN STUDIES
THE UNIVERSITY OF MICHIGAN   NUMBER 32

*Open access edition funded by the National Endowment for the Humanities/
Andrew W. Mellon Foundation Humanities Open Book Program.*

Library of Congress Catalogue Card Number: 87–62020

ISBN: 0–89148–043–9 (cloth)
ISBN: 0–89148–044–7 (paper)

Copyright © 1988

Center for South and Southeast Asian Studies
The University of Michigan

ISBN 978-0-89148-043-3 (hardcover)
ISBN 978-0-89148-044-0 (paper)
ISBN 978-0-472-12830-3 (ebook)
ISBN 978-0-472-90230-9 (open access)

To the memory of
Dawn Jean Rambo

CONTENTS

Contents

MAPS, FIGURES, AND TABLES

Maps

Figures

Tables

PREFACE

In August 1984, a three-day conference on ethnic diversity and the control of natural resources was held at the University of Michigan. Jointly sponsored by the Center for South and Southeast Asian Studies of the University of Michigan and the Environment and Policy Institute (EAPI) of the East-West Center in Honolulu, the conference was attended by fifteen scholars from six Asian and Pacific countries including the United States (see list of participants in the Appendix). This conference was the second in a series of conferences dealing with key issues in the human ecology of Southeast Asia jointly sponsored by the East-West Center and the University of Michigan. The proceedings of the first conference, held in Honolulu in 1983, have been published in *Cultural Values and Human Ecology in Southeast Asia* as Number 27 in the series Michigan Papers on South and Southeast Asia.

Like the first conference, which focused on the relationship between cultural values and tropical ecology, this meeting was planned to bring together Asian scholars working in member projects of the Southeast Asian Universities Agroecosystem Network (SUAN) with scholars in Western institutions who share their concern with developing human ecology research in the region. The purpose of the conference was to explore the ecological ramifications of ethnicity in Southeast Asia. This topic was chosen under the assumption that the high degree of ethnic diversity historically evident in the region and its continuing importance under contemporary conditions is related to both the nature of tropical Asian environments on the one hand, and the nature of Southeast Asian political systems and the ways in which they manipulate natural resources on the other. Although the emphasis of the conference series is on developing improved theoretical understandings of human-environment interactions, it should be evident that the topics selected for discussion also have great practical significance.

Scholars representing a broad range of theoretical perspectives and working with an even wider range of ethnic groups and environments were invited. This diversity is reflected in the resulting papers although, unfortunately, the range is narrower than it was in the conference itself. This reflects the fact that a number of participants, for a variety of reasons outside the control of the editors, elected not to submit their papers for publication in this volume.

All the chapters have been significantly revised following the conference, in many cases taking alternative perspectives presented by other participants into account in the process. The editors have sought to impose a common format while trying to preserve as much as possible the unique qualities of the individual papers.

Funding for the conference was provided by the East-West Center and the University of Michigan. Karen Ashitomi and Regina Gregory, EAPI Human Ecology Program secretaries, typed multiple drafts of the manuscripts. Helen Takeuchi was responsible for copyediting.

CHAPTER 1

INTRODUCTION

A. Terry Rambo
Karl L. Hutterer
Kathleen Gillogly

Its vast human diversity—racial, linguistic, and cultural—has made Southeast Asia simultaneously the ethnographer's heaven and the ethnologist's purgatory. Simply describing the multitude of different peoples inhabiting the region has kept several generations of field workers gainfully employed, with results that threaten to swamp all but the largest library collections. As a consequence of this activity, it is fair to say that Southeast Asia is now at least as well known ethnographically as any world region with the exception of Native North America.

Unfortunately, ethnologists have not enjoyed great success in imposing classificatory order on the diversity revealed by the work of the ethnographers. The culture-area approach, still the standard framework for ordering New World ethnology, was not a conspicuous success when applied to Southeast Asia (Bacon 1946; Kroeber 1947; Narroll 1950). To dichotomize the multitudinous groups in the area into upland swidden farming tribesmen and lowland wet rice farming peasants (Burling 1965) radically oversimplifies reality. Attempts to retain complexity, as in the ethnolinguistic classification employed by LeBar, Hickey, and Musgrave (1964), lack any systematic basis for differentiating between groups. As Yengoyan points out in his chapter in this volume, ethnographers have tended to "divide and subdivide groups into finer categories whose existence is highly suspect."

The ad hoc character of ethnic classifications in Southeast Asia reflects, in our view, the continuing theoretical underdevelopment of area studies. Rather than evolving out of a vigorous confrontation between ethnographic observations and general anthropological theory, as was the case in New World ethnological classifications, the schemes employed in Southeast Asia have usually been imported ones. Unlike the American culture-area classification schemes, which are based upon a complex set of theoretical explanations for ethnic differentiation (adaptation to common environmental factors within each area plus diffusion across area boundaries), most of the schemes employed in Southeast Asia have assumed that diversity is explainable in terms of a single cause. For a long time, migration and diffusion were invoked as the key explanatory factor; more recently, equally simplistic evolutionary models have begun to displace the culture-historical approach.

Waves of Migration

In the "migratory waves" model, existing cultural groups in Southeast Asia were assumed to be the residues of a series of discrete migrations into the region. Each successive wave of culturally more advanced people was supposed to have pushed the more backward earlier arrivals deeper into mountain and forest refuges where they had maintained their culture in pristine condition awaiting the coming of European ethnologists. Thus the scattered nomadic forest foraging groups such as the Aeta (Agta, Ata) of the Philippines and the Semang of Peninsular Malaysia were assumed, because of cultural "primitiveness," to be survivors of the earliest migratory wave. The numerous different tribal groups practicing swidden agriculture (e.g., the Ilongots of Luzon and the Senoi Semai of Peninsular Malaysia) represented a subsequent set of waves often referred to collectively as "Proto-Indonesians," while the dominant national lowland populations such as the Khmer, Thai, and Malay were seen as representing more recent migrations into the region.

Fay-Cooper Cole's *The Peoples of Malaysia* (1945), for two decades the standard English language ethnology for insular Southeast Asia, exemplifies the migratory wave model. With minor modifications the same conceptual framework is employed in several recent regional ethnologies (Burling 1965; Carey 1976; Wallace 1971). This is quite surprising when it is remembered this model relies on a view of culture history that was formulated early in the twentieth century in Germany by Graebner (1905, 1911) and Schmidt (1906a, 1911) and the followers of the *Kulturkreislehre* and specifically adapted to Southeast Asian culture history, first on the basis of linguistic arguments by Schmidt (1906b, 1926), and only shortly later on the basis of archaeological and ethnographic arguments by Heine-Geldern (1923, 1932, 1937, 1946), Beyer (1932, 1948), and Schebesta (e.g., 1927, 1952-57). A slightly different and even more extreme version was proposed by the physical anthropologist von Eickstedt (1944). The model is based upon the assumption that racial, linguistic, and cultural traits comprise stable packages that persist unchanged for millennia.

> The transition from the Palaeolithic to the incipient Neolithic of
> Tongkin and the Malay Peninsula occurred gradually and
> probably did not involve major modifications of the cultures in
> question. This changes with the beginning of the full Neolithic.
> New peoples and cultures pour in in force from the north and
> northeast. With this, the foundations are laid, clearly discer-
> nible today, for all subsequent historical developments and for
> the present distribution of races, peoples, languages, and
> cultures. (Heine-Geldern 1932: 556—translation by
> K.L. Hutterer)

The durability of this model is all the more surprising given the fact that Boas
and other American ethnologists had already shown by the early 1920s that its
underlying assumptions are contrary to the empirical evidence. The persistence
of such a theoretically dubious model for such an extended period, despite its
general rejection elsewhere in the world, raises troubling questions about the
state of ethnological research in Southeast Asia. One must wonder why new
anthropological concepts have taken so long to penetrate into the thinking of
"area study" specialists and why the serious problems they encountered in fitting
empirical data into the traditional framework did not lead much sooner than it
did to the search for new theories.

Evolutionary Models of Ethnic
Diversification in Southeast Asia

In the late 1960s and early 1970s a new conceptual framework for Southeast
Asian culture history challenged the long dominant migratory waves model.
Primarily the product of Wilhelm Solheim and his graduate students (e.g.,
Chester Gorman, Donn Bayard) at the University of Hawaii, the new model
shifted emphasis from migration and diffusion to evolution in place. Claims,
unfortunately poorly supported by hard evidence, were made that agriculture,
pottery making, and metallurgy had all originated in Southeast Asia centuries
and even millennia before their "discovery" in China and the Middle East (e.g.,
Bayard 1970, 1979, 1980; Gorman 1969, 1977; Solheim 1969, 1970, 1975, 1979).
Southeast Asia was suddenly transformed from the trash can of history to the
primal source of civilization. If any waves were perceived by advocates of the
new model, they were seen as flowing outward to the rest of the world, bearing on
their crest the seeds of future progress.

The new evolutionary framework gained rapid acceptance among both
Western scholars and educated elites in the Southeast Asian countries. In the
latter case one may speculate that the new model was compatible with the

heightened sense of nationalism that accompanied the liberation of most of the region from colonial rule. The Vietnamese, for example, freed from using French history texts that declaimed about "our ancestors the Gauls" were hardly eager to have to recite about "our ancestors the Chinese" instead. It was far more satisfying to accept Gorman's thesis that agriculture had originated among the Hoabinhian people of northern Indochina and had spread northward from them to the Chinese. The last had indeed become first.

As might be expected, the situation among professional scholars was somewhat more complex. Certainly, the academically marginal status of Southeast Asian studies in contrast to South and particularly East Asian studies was a sore spot with many. There is no denying the sense of satisfaction that many of us felt when we first heard the new claims for Southeast Asia as an evolutionary hearth for world civilization. At the same time, the ground had already been prepared for such a "paradigm shift" by the increasing discordance between empirical research findings on ethnic groups in the region and the theoretical assumptions underlying the migratory waves model (cf. Bronson 1977). Many anthropologists had read and been disquieted by Leach's *Political Systems of Highland Burma* (1954). If Kachin could transmute themselves into Shan, what did that do to the assumption of ethnic constancy that was required by the migratory waves model? An evolutionary explanation of the origins of ethnic diversity seemed to offer one viable alternative. Curiously, two quite different evolutionary models were proposed almost simultaneously: one reflecting concern with ecological adaptation and the other reflecting concern with inter-ethnic structural relations.

Ecological Adaptation

Attempts to explain ethnic diversity as the result of ecological adaptation reflected the growing recognition in the 1950s that cultural traits, just as much as racial characteristics, were subject to natural selection. The acceptability of this approach was increased by the vogue adaptionist explanations then enjoyed in general anthropological theory. In physical anthropology, many assumed markers of racial relationship (e.g., body stature, skin color, even abnormal hemoglobins) were shown to be subject to natural selection. Proponents of cultural ecology advanced similar claims regarding many cultural traits that had previously been considered markers of ethnic phylogeny. From this evolutionary perspective, ethnic diversity is explained in terms of adaptive radiation in which cultural differentiation arises as groups confront distinctive, locally specific environmental selection pressures. For example, Rambo (1973) has shown how Southern Vietnamese peasant society became differentiated from its ancestral

Northern form in the course of becoming adapted to the distinctive new environment of the Mekong Delta.

Although earlier studies tended to emphasize the adaptation of culture to the natural environment, it has been increasingly recognized that groups must also adapt to the social field in which they are located. Lehman (1963), for example, has suggested that the structure of Chin society in highland Burma reflects an adaptation to the lowland Burmese state as much or more as it does the influence of the natural environment. Similarly, Hutterer (1974) has proposed that Philippine lowland societies evolved as a social adaptation to the demands and opportunities of maritime trade with societies on the Asian mainland.

The adaptionist stance is well represented in this volume. Rambo's chapter on Malayan aboriginal ethnogenesis offers an admittedly "just so story" hypothetical reconstruction of how racially and culturally distinct Semang hunters and gatherers and Senoi swidden agriculturalists could have evolved in place out of a common ancestral population as the result of adaptation to different selective pressures associated with the distinct niches filled by these two groups. Eder, Gomes, Pei, and Winzeler also describe cases in which ethnic differentiation reflects ecological adaptation, but place much greater emphasis on the importance of selective pressures exercised by the larger social environment. Gomes and Winzeler stress the role of political forces associated with modern nation-states in the formation and maintenance of ethnic groups. Pei and Eder emphasize adaptation to the economic environment created by the existence of intergroup trade. In focusing on the adaptation of ethnic groups to their social environment, these studies converge on the central concern of the structural approach to an evolutionary interpretation of ethnic diversity—the model of intergroup interactions.

Intergroup Interactions

An alternative evolutionary model for explaining ethnic diversity is the transactional approach associated with Fredrik Barth. Barth sees differentiation as emerging from interactions between neighboring groups competing for limited resources. The stronger the potential competition, the more pronounced the degree of differentiation. To use ecological terminology, the alternative to competitive exclusion is character displacement, with each group coming to occupy a unique niche that provides it with exclusive access to the resources necessary for its survival. Individuals choose to manifest the appropriate cultural markers for a particular group in order to claim the resources assigned to its niche (see Foster, this volume). In this way, the ethnic groups partition their environment and increase their degree of specialization in resource exploitation, thereby increasing each group's energetic efficiency. Differences of emphasis are possible within this approach. Abruzzi (1982) has moved this approach in the direction of the adaptionist stance by arguing that speciation and ethnic differentiation can be explained in terms of the same ecological process. Benjamin (1985), on the other hand, has stressed the importance of deliberately maintained symbolic traditions.

The growing popularity among anthropologists of ways of thinking about culture in terms of the structural properties of relations between social groups rather than in terms of "shared culture" (see Foster, this volume) contributed to the acceptance of the intergroup interaction model. Cultural traits were seen as essentially epiphenomenal indicators of underlying structural relationships between social formations, subject to rapid change as these relationships shifted over time. In the most extreme formulations, ethnic diversity was seen as the product of the expansion of capitalism into the peripheral regions of Asia, Africa, and the Americas. Eric Wolf, the leading proponent of this view among anthropologists, asserts that

> anthropologists look for pristine replicas of the precapitalist, preindustrial past in the sinks and margins of the capitalist, industrial world. But Europeans and Americans would never have encountered these supposed bearers of a pristine past if they had not encountered one another, in bloody fact, as Europe reached out to seize the resources and populations of the other continents. Thus, it has been rightly said that anthropology is an offspring of imperialism. Without imperialism there would be no anthropologists, but there would also be no Dene, Baluba, or Malay fishermen to be studied. (Wolf 1982:18)

Wolf's claim is clearly hyperbole. The Malay fishermen to which he refers had established a distinct ethnic identity (although not, perhaps, the one that Raymond Firth encountered when he did his fieldwork in Kelantan in the 1940s) in their interactions with the Chinese, Thai, and the aboriginal groups of the interior forests long before the Portugese, Dutch, and English arrived on the scene. One need not, however, lay total blame on European capitalism to accept the view that cultures evolve as much in the context of a political, economic, and social environment as they do a biophysical one and that, consequently, structural relations between groups are a significant factor in the generation of ethnic diversity.

Many of the chapters in this volume are written from such an intergroup-interaction perspective. They are written within the tradition of Barth but go beyond his formulations. Leach (1954) is cited nearly as often as Barth and, in the case of Yengoyan's analysis of Mandaya ethnicity, is a central part of the argument. Eder and Winzeler, while working out of a basically adaptationist framework, also pay attention to the influence of external economic and political power on local ethnic relations. The hegemony that immigrant lowlanders exercise over tribal groups is considered by both Eder and Yengoyan: Eder emphasizes economic hegemony and Yengoyan stresses political and symbolic hegemony, as does Wee from a quite different perspective.

Rosaldo, Wee, and Yengoyan consider how various expressions of ethnicity relate to concepts of hierarchy and the social structural positions assumed when thinking of ethnic identifications. Besides the pervading theme of the relationship between status stratification and ethnicity, there is a concomitant theme of interior/exterior divisions (Yengoyan), or a "circle of belonging" (Wee) associated with egalitarianism and cohesiveness of those who belong inside. Rosaldo suggests that ethnicity among the Ilongots is simultaneously a strategic presentation of their identity in contrast to neighboring groups and a fundamental aspect of their sense of cultural existence. They are "Ilongot" on the boundaries of their territory because they are in confrontation with members of different groups but they are "Ilongot" in the interior of their territory because they are participants in the ongoing cultural traditions of their society.

Cultural History and Evolution as Complementary Explanations of Ethnic Diversity in Southeast Asia

Proponents of the various theoretical approaches for explaining ethnic diversity have usually presented their respective models as if they were mutually exclusive explanations. If one approach was correct, then others were necessarily invalid. The chapters in this volume, however, suggest that no single process can satisfactorily account for the ethnic diversity observed in Southeast Asia. Instead, multiple processes have been involved.

Those readers who, like the editors, had largely rejected the migratory waves model in favor of an evolutionary explanation are likely to be surprised by the strength of the evidence presented in the chapters in this volume for the importance of migration as a source of ethnic diversity in the region. In some cases, such as the movement of Cebuanos into the territory formerly occupied by the Ata and Batak described by Cadeliña and Eder, and that of the Han Chinese into the Dai areas of Xishuangbanna described by Pei, these migrations are still in process. In other cases, such as the Bugis migration into the Riau Islands described by Wee and the settlement of the Thai and Chinese in Kelantan described by Winzeler, the process was completed before the imposition of European colonial rule.

Clearly, these case studies confront us with strong evidence that migration has contributed in a major way to the human diversity of the region. The role of migration as a source of ethnic diversity is compatible with the emphasis on geographical isolation in current theories of speciation in evolutionary biology. Although the possibility of sympatric speciation under special conditions is still contested, it is generally accepted that allopatric speciation is by far the more common path to biological diversity (Mayr 1970).

Also significant is that, in many of the cases described in this volume, populations associated with more complex chiefdom or state level social formations have intruded into the territories of simpler foraging and swidden farming groups, a finding that is also in accord with the migratory wave model. Presumably, these more complex societies have greater military strength and more efficient organizational and administrative institutions, which give them the capability to take over territory belonging to weaker bands and tribes.

The pattern of migration reported in these chapters differs from that postulated by the traditional migratory waves model, however, in that in each specific locality we are dealing with different "advanced" groups—Cebuanos in the case of the Ata and Batak, Bugis in the case of the Orang Laut, Malays and Chinese in the case of the Semai—rather than a single unitary wave sweeping over the entire region. Thus, migration has contributed to the observed ethnic diversity of Southeast Asia, but not in the simple layer-cake manner envisioned in the grand culture-historical schemes of Heine-Geldern and Beyer.

Migration is responsible for the presence within the same area of initially distinctive groups having different potentials to control and exploit resources, but the subsequent interaction between these groups shapes their respective ethnic identities. As Eder suggests, there is a profound difference between the character of hunting and gathering societies living in pristine subsistence situations and those, such as the Agta, who live by exchanging meat for grain with agricultural groups. In the latter case the necessity of maintaining trading relationships seems to determine the value orientations of the Agta (e.g., reluctance to engage in agriculture), which other observers have seen as survivals of their traditional culture. In this regard, then, the interactionist model appears to be valid. Ethnic identity, rather than reflecting the persistence of historical cultural traditions, is the constantly changing product of contemporary relationships between social groups.

One important limitation on the malleability of ethnic identity markers, which is ignored in most of the literature employing the intergroup-interaction model, is referred to in several of the chapters, however. That is, the existence of directly observable, genetically controlled "racial" characteristics (e.g., skin color, stature, hair form), unlike cultural or linguistic markers, cannot be changed at individual volition. The Ata and Batak described by Cadeliña and Eder cannot simply choose to become Cebuano in the same way that the Kachin can become Shan, or the Mon described by Foster can become Thai. In such cases, which are far from unusual in Southeast Asia, ethnic distinctiveness may persist in situations that would result in merging of groups in the absence of physical differences. Thus, in the Ata case described by Cadeliña, even the adoption of a mode of subsistence essentially identical to that of their Cebuano neighbors has not resulted in the disappearance of the Negritos as a distinct ethnic group.

Yengoyan describes a somewhat different case of physical characteristics preventing a switch in ethnic identity: individual Mandaya cannot successfully become lowland Filipinos because they are unable to conceal their tattoos. In this case, the maintenance of a cultural practice resulting in distinctive physical characteristics acts to preserve ethnic distinctiveness and identity. Abandoning this practice by the group as a whole would not only make acculturation possible but would, in itself, signal loss or abandonment of a cultural tradition that is part and parcel of the group's identity. This theme of deculturation rather than acculturation as the outcome of interactions with other ethnic groups is also considered in Rosaldo's chapter.

Another important constraint on the extent to which ethnic identity is determined by intergroup interactions—particularly assimilation of subordinate groups into groups associated with dominant "national" cultures—is the growing power of states to jurally regulate ethnic group membership. As Winzeler shows in his chapter, even if the Chinese and Thai minorities in Kelantan should want to "become" Malay—a change in ethnicity that would greatly enhance their access to land and other resources—they are effectively barred from doing so by Malaysia's national legal code. Although according to local custom non-Malays can "become" Malay by changing their religion to Islam and their language to Bahasa Melayu, and by displaying appropriate cultural behavior in public, only individuals born to Malay parents are legally allowed to purchase land within the Malay Reserve areas, which comprise the bulk of the arable land in the state. The marking of ethnic group membership on national identity cards, as is done in Malaysia and several other modern Southeast Asian nations, may be far more effective than racial features or tattoos in freezing the current ethnic diversity of the region within its present set of categories.

Both the ecological adaptation and the intergroup interaction models emphasize the importance of differential access to natural resources in generating and maintaining ethnic diversity. This emphasis on control of resources is well supported by several chapters in this volume. In particular, the expansion of groups specializing in full-time market-oriented agricultural production into tropical forest areas seems to create a niche for groups specializing in collection of forest products to trade with the agriculturalists. The Semang in Malaysia described by Rambo and the Agta and Batak in the Philippines described by Eder are examples of this creation of what Richard Fox (1969) labeled "professional primitives." The knowledge, skills, and value orientations required to exploit forest resources on the one hand and manage intensive agricultural production on the other are so different that they require mutually exclusive economic specializations. In relatively large and complex agricultural societies, it may be possible to maintain forest specialists within the same cultural system. More commonly, however, the demand for forest products provides an opportunity for economic specialization along ethnic lines. As long as agriculture does not expand to the point where it destroys all the forest, a secure niche remains for the "hunters and gatherers." In fact, as Eder suggests, their conversion to agriculturalists, despite abundant opportunities for knowledge of farming to diffuse to them, is inhibited by their ecological specialization as forest collectors as well as by lowlander hegemony.

The significance of differential control of natural resources is more problematic in cases where all the groups in an area are involved in agriculture. Ecological specialization may still play a role in ethnic survival as in the case of the Kelantan Chinese described by Winzeler. Inhabiting sandy, well-drained river banks that are unsuited for rice farming practiced by the Malays, these Chinese have specialized in vegetable production. Rather than being competitive, the productive activities of the two groups are complementary. In many other cases, however, expansion of commercial agriculture by one ethnic group results in the competitive exclusion of subsistence agriculturalists belonging to other ethnic groups. This appears to be the usual fate of swidden farming "tribes" on the margins of capitalist agricultural expansion, such as the Mandaya described by Yengoyan. Such groups must either retreat into remote refuge areas unsuitable for intensive agricultural development or they gradually lose their ethnic distinctiveness as they are absorbed into the wage labor proletariat employed on the plantations.

In an ecologically homogeneous agricultural area, however, cultural factors that influence the ability of different groups to exploit resources can, under some circumstances, provide in themselves a basis for maintenance of ethnic diversity. Thus, the Thai in Kelantan described by Winzeler raise pigs in their villages and hunt wild pigs that invade their farms from the forest. The latter activity benefits their Malay neighbors who as devout Muslims are forbidden to come into close contact with pigs. In effect, culture creates the differentiation in access to natural resources necessary for the survival of Thai and Malay farmers as separate ethnic groups. It is probably more often the case, at least in contemporary Southeast Asia, that cultural difference is itself the resource that justifies maintenance of distinctive ethnic groups. Foster describes this kind of situation about the Mon in modern Thailand.

According to Foster, Mon migrants from Burma established numerous rice farming villages in Central Thailand in the eighteenth and nineteenth centuries. Over the past century acculturation has been rapid, and most of the descendants of these migrants have been assimilated into the numerically and politically dominant rural Thai population. Today, the only persons who still assert Mon ethnic identity are those few individuals engaging in commerce, especially the trade in household pottery. Foster explains the persistence of Mon identity among the latter group in terms of its serving to differentiate them from the Thai peasants with whom they trade. By making them "outsiders" in the village society, having Mon identity frees the traders from "the traditional expectations of generosity and fair dealing which are so contrary to the viable pursuit of trade."

The Mon traders gain this economic advantage at the cost of retaining an ethnic identity held in low esteem by their Thai neighbors. Thus relations between groups in the "symbolic economy" may be an inversion of their relationships in the material economy. The rich "outsider" merchant is considered socially inferior to the poor "insider" farmer. Such inversions, as Vivienne Wee's elegant analysis of the material and symbolic aspects of ethnic diversity in the Riau Islands suggests, may be a common characteristic of ethnically complex state societies in the region. According to Wee, the groups that are most tightly integrated into the national Indonesian economy and are thus the most dependent upon powerful external forces, which they are too weak to control, are at the top of the ethnic status hierarchy in the Riaus. The subsistence-oriented Orang Laut, on the other hand, who are in material terms the most independent group, are symbolically classed in the lowest and most dependent category. The symbolic aspects of ethnicity serve to conceal the harsh realities of class dependency within the political economy of Indonesia. (The Orang Laut, however, hold to an alternative symbolic economy that portrays them as the true "indigenes" while the aristocrats of the dominant system are seen as "foreigners." The Orang Laut therefore acknowledge an inclusive/exclusive social division rather than accept a status hierarchy that denigrates them. Both formulations serve to obscure political and economic realities.)

In complex, class-stratified states, the ability to manipulate ethnic consciousness among the lower classes may become a key resource for the national elites in maintaining their political and economic dominance. The "New Economic Policy" adopted in Malaysia following the communal rioting in 1969 represents an official commitment to eradicate Malay poverty by forcing redistribution of control over the economy so as to reduce inequalities among the shares held by the several ethnic groups. As interpreted by the current Prime Minister, however, success will have been achieved when there are as many Malay millionaires as there are wealthy Chinese. The votes of poor rural Malays, mobilized in support of the government by appeals to ethnic solidarity, provide the political power neccessary to increase the wealth of the Malay elite. At the same time, ethnic chauvinism effectively inhibits the formation of multiracial, class-based interest groups such as trade unions that might advocate more egalitarian redistribution of wealth and power.

Conclusion

The preceding discussion shows no single existing conceptual approach is adequate to the task of explaining the ethnic diversity of Southeast Asia. Each accounts for some of the diversity but none is sufficiently comprehensive to explain the total range of ethnic phenomena observed in the region.

Previous commentators (e.g., Despres 1975; Cohen 1978) have pointed to the lack of a consistent definition of ethnicity as the major obstacle in the way of successful comprehensive comparative analysis. Definitional problems do not appear, however, to have been a major constraint on the contributors to this volume; none of them seem to have encountered any difficulty in working with what is, admittedly, a rather loosely defined concept. This may be a reflection of the fact that all the groups they studied also employ some such concept within their daily lives. These folk concepts of ethnicity are not all identically defined but all fit comfortably inside the boundaries that scientists assign to the term.

Rather than focus on defining the phenomenon of ethnicity, it seems to us more fruitful to examine the different social evolutionary contexts in which the phenomenon is manifested. Thus, rather than expecting to discover a single, universally valid explanation for the existence of ethnic diversity, we suspect different explanations are applicable to different social evolutionary stages. In a world entirely composed of bands or tribes, adaptive radiation may be the principal driving force behind ethnic diversification. With the emergence of complex multigroup societies, bringing groups with very different resource exploitation patterns into structured relations, intergroup interactions may become more significant in defining ethnicity. Finally, in modern nation-states, where ethnic group membership is legally regulated by the central government, the legislative process itself becomes the key factor in ethnic relations.

Such an evolutionary pattern seems to be suggested by the chapter contents of Rambo, Gomes, and Winzeler on Malaysia, and by Rosaldo, Eder, Cadeliña, and Yengoyan on the Philippines. In each case initial differentiation appears to have been the result of adaptation to distinct niches, as in the separation of the Semang from the Senoi as the former became specialized forest collectors and the latter swidden farmers as hypothesized by Rambo. The formation of complex, multigroup social systems (in which migration often plays a critical role in creating initial diversity) gives a greater significance to interactions between groups in shaping ethnic identity, as described in the Agta-lowlander nexus by Eder and the Semai-Malay nexus by Gomes. Incorporation of control over ethnicity within the jural authority of the nation-state, as in Malaysia's constitutional enshrinement of criteria for becoming Malay described by Winzeler, marks a reversal in the causal connection between ethnic identification and control over resources from that which prevailed in earlier evolutionary stages. Whereas in the earliest stage ethnic divisions reflected the existence of differing adaptations to distinct niches, and then, as the social system grew in complexity, came to symbolically reinforce the social division of labor built on such adaptive differences, in the nation-state context ethnic identity becomes a powerful tool for expanding a favored group's control over resources. Thus, to put it in admittedly overly simple terms, in the premodern Malay Peninsula people became Malay because they were wet-rice farmers, but in the modern state of Malaysia they become civil servants, army officers, and businessmen because they are Malay.

Malaysia unquestionably represents the extreme case in the region of state-controlled imposition over ethnic relations. In Thailand, Mons can still freely choose to become Thai, as Foster reports (although it is much less certain that the Cambodian and Laotian refugees in the border camps will be given the same opportunity); and in the Philippines the Ata and the Mandaya described by Cadeliña and Yengoyan can legally become Cebuano, although such conversion is inhibited by their distinctive physical characteristics (or, in the Mandaya case, by tattoos), which visibly distinguish them from the dominant lowlanders. The overall trend within Southeast Asia, however, appears to be in the direction taken by Malaysia, with opportunities for ethnic change increasingly circumscribed by legal restrictions. The maintenance of such strict boundaries is possible in part through improved surveillance techniques (such as identity cards and documentation of individuals' parentage) found in modern nation-states.

The evolutionary sequence we have sketched here is only a tentative attempt to develop a more comprehensive conceptual approach to the explanation of ethnic diversity in Southeast Asia. The three stages we propose are admittedly hypothetical ones, based more on the logic of social evolutionary theory than on full empirical analysis. We suspect that a much more complicated scheme will be needed to incorporate the different historical trajectories followed by different parts of the region.

Thus, although we are convinced that an evolutionary perspective of the sort we have proposed is useful, we are under no illusion that it represents the grand theoretical synthesis that definitively solves all the problems of Southeast Asian ethnology. As the chapters in this volume reveal, the empirical realities are so incredibly diverse and complex that no single conceptual framework, however broad, appears able to encompass every significant feature. We are all, as Vivienne Wee points out, adherents of theories that are "hocus-pocus" rather than "God's truth." Thus our advocacy of an evolutionary approach to the understanding of ethnic diversity does not imply that it is "the only valid approach or that the same ethnographic data may not be analyzed in some other ways." Instead, we believe that theoretical pluralism offers the best hope of ultimately generating the conceptual frameworks still needed to order in full the kinds of diverse observational data arrayed in this volume.

References

Abruzzi, W.S.
    1982 Ecological theory and ethnic differentiation among human popula-
        tions. *Current Anthropology* 23(1):13–35.

Bacon, Elizabeth
    1946 A preliminary attempt to determine the cultural areas of Asia.
        Southwestern Journal of Anthropology 2(2):117–132.

Bayard, Donn T.
    1970 Excavations at Non Nok Tha, northeastern Thailand, 1968. *Asian
        Perspectives* 13:109–143.
    1979 The chronology of prehistoric metallurgy in north-east Thailand:
        Silabhumi or Samrddhabhumi? In *Early South East Asia*, edited
        by R.B. Smith and W. Watson. New York: Oxford University
        Press. Pp. 15–32.
    1980 An early indigenous bronze technology in northeast Thailand: Its
        implications for the prehistory of East Asia. In *The Diffusion of
        Material Culture*, edited by H.H.E. Loofs-Wissowa. Honolulu:
        University of Hawaii Social Science Research Institute. Pp. 191–
        214.

Benjamin, Geoffrey
    1985 In the long term: Three themes in Malayan cultural ecology. In
        Cultural Values and Human Ecology in Southeast Asia, edited by
        Karl L. Hutterer, A. Terry Rambo, and George Lovelace. Ann
        Arbor: University of Michigan Center for South and Southeast
        Asian Studies, Paper No. 27. Pp. 219–278.

Beyer, H. Otley
    1932 A tabular history of the Philippine population, as known at the
        present time from combined historical, ethnographical, and
        archaeological studies. In *Praehistorica Asiae Orientalis, Premier
        Congres des Prehistoriens d'Extreme-Orient*, Vol. 1. Hanoi:
        Imprimerie d'Extreme-Orient. Pp. 129–135.
    1948 Philippine and East Asian archaeology and its relation to the origin
        of the Pacific Islands population. *Bulletin of the National Research
        Council of the Philippines* 29:1–130.

Bronson, Bennett
    1977 Against migration: A negative perspective on population move-
        ments in prehistoric Southeast Asia. *Kabar Seberang* 1:29-43.

Burling, Robbins
    1965 *Hill Farms and Padi Fields: Life in Mainland Southeast Asia.*
        Englewood Cliffs, NJ: Prentice-Hall.

Carey, Iskander
    1976 *Orang Asli: The Aboriginal Tribes of Peninsular Malay-sia.* Kuala
        Lumpur: Oxford University Press.

Cohen, Ronald
    1978 Ethnicity: Problems and focus in anthropology. *Annual Review of
        Anthropology* 7:379-403.

Cole, Fay-Cooper
    1945 *The Peoples of Malaysia.* Princeton, NJ: Van Nostrand.

Despres, L.S.
    1975 *Ethnicity and Resource Competition in Plural Societies.* The
        Hague: Mouton.

Fox, Richard G.
    1969 Professional primitives: Hunters and gatherers of nuclear South
        Asia. *Man in India* 48:139-160.

Gorman, Chester F.
    1969 Hoabinhian: A pebble-tool complex with early plant associations in
        Southeast Asia. *Science* 163:671-673.
    1977 A priori models and Thai prehistory: A reconsideration of the
        beginnings of agriculture in Southeast Asia. In *Origins of Agricul-
        ture*, edited by C.A. Reed. The Hague: Mouton. Pp. 321-355.

Graebner, Fritz
    1905 Kulturkreise und Kulturschichten in Ozeanien. *Zeitschrift für
        Ethnologie* 37:28-53.
    1911 *Die Methode der Ethnologie.* Heidelberg: Karl Winter's Univer-
        sitätsbuchhandlung.

Heine-Geldern, Robert
    1923 Südostasien. In *Illustrierte Völkerkunde*, edited by Georg
        Buschan. Stuttgart: Strecker und Schroeder. Pp. 689-968.
    1932 Urheimat und früheste Wanderungen der Austronesier. *Anthropos*
        27:543-619.

16 *Rambo, Hutterer, Gillogly*

1937 L'art prebouddhique de la China et de l'Asie du Sud-est et son influence en Oceanie. *Revue des Arts Asiatiques*, Vol. 11.
1946 Research on Southeast Asia: Problems and suggestions. *American Anthropologist* 48:149–175.

Hutterer, Karl L.
1974 The evolution of Philippine lowland societies. *Mankind* 9:287–299.

Kroeber, A.L.
1947 Culture groupings in Asia. *Southwestern Journal of Anthropology* 3(4):322–330.

Leach, Edmund
1954 *Political Systems of Highland Burma.* Cambridge: Harvard University Press.

LeBar, F.M., G.C. Hickey, and J.K. Musgrave
1964 *Ethnic Groups of Mainland Southeast Asia.* New Haven, CT: Human Relations Area Files Press.

Lehman, F.K.
1963 *The Structure of Chin Society: A Tribal People of Burma Adapted to a Non-Western Civilization.* Illinois Studies in Anthropology No. 3. Urbana, IL: University of Illinois Press.

Mayr, Ernst
1970 *Population, Species, and Evolution.* Cambridge, MA: Belknap Press.

Narroll, Raoul S.
1950 A draft map of the culture areas of Asia. *Southwestern Journal of Anthropology* 6(2):183–187.

Rambo, A. Terry
1973 *A Comparison of Peasant Social Systems of Northern and Southern Viet-Nam.* Monograph Series 3. Carbondale, IL: Southern Illinois University Center for Vietnamese Studies.

Schebesta, Paul
1927 *Among the Forest Dwarfs of Malaya.* London: Hutchinson.
1952–57 *Die Negrito Asiens.* 2 vols. Wien-Mödling: St. Gabriel Verlag.

Schmidt, Wilhelm

   1906a Die moderne Ethnologie. *Anthropos* 1:134–163, 318–387, 593–643, 950–997.

   1906b *Die Mon-Khmer Völker, ein Bindeglied zwischen den Völkern Zentralasiens und Austronesiens.* Braunschweig: F. Viehweg.

   1911 Die kulturhistorische Methode in der Ethnologie. *Anthropos* 6:1010–1036.

   1926 *Die Sprachfamilien und Sprachkreise der Erde.* Heidelberg: Kulturgeschichtliche Bibliothek.

Solheim, Wilhelm G., II

   1969 Reworking Southeast Asian prehistory. *Paideuma* 15:125–139.

   1970 Northern Thailand, Southeast Asia, and world prehistory. *Asian Perspectives* 13:145–162.

   1975 Reflections on the new data of Southeast Asian prehistory: Austronesian origin and consequence. *Asian Perspectives* 18:146–160.

   1979 A look at "L'art prebouddhique de la Chine et de l'Asie du Sud-est et son influence en Oceanie" forty years after. *Asian Perspectives* 22:165–205.

von Eickstedt, Egon Frhr.

   1944 *Rassendynamik von Ostasien.* Berlin: Walter de Gruyter.

Wallace, Ben J.

   1971 *Village Life in Insular Southeast Asia.* Boston, MA: Little, Brown and Company.

Wolf, Eric R.

   1982 *Europe and the People Without History.* Berkeley, CA: University of California Press.

CHAPTER 2

WHY ARE THE SEMANG?
ECOLOGY AND ETHNOGENESIS OF ABORIGINAL
GROUPS IN PENINSULAR MALAYSIA

A. Terry Rambo

As the British gradually extended their colonial rule over the Malay peninsula, they came to perceive the indigenous population as being divided into two
very different ethnic groups. On the one hand were the Muslim Malays who
were perceived by the British as belonging to the "civilized races." On the other
hand were the "pagan" aborigines who were perceived as the disappearing relics
of the primitive races that had once been the sole inhabitants of the peninsula.
Initially, the aborigines were referred to by a diverse and confusing collection of
generic Malay names, e.g., *sakai* (subject or slave), *orang liar* (wild people), *orang
ulu* (people of the headwaters), and *orang hutan* (people of the forest). Specific
groups were referred to by an even more bewildering collection of terms based on
local place names, often "people of the *x* valley or *y* river," or on ethnic labels
used by neighboring groups. The complexities of Malayan aboriginal ethnonyms
are discussed in greater detail by Gomes in Chapter 6 of this volume.

It was not until the turn of the twentieth century that classificatory order
was first imposed upon this ethnographic chaos. Skeat and Blagden, in their
monumental *Pagan Races of the Malay Peninsula* (1906), suggested that the
aborigines could be divided into three basic ethnic groups: Semang, Sakai, and
Jakun. In an article published in 1926, Father Paul Schebesta further codified
this classificatory scheme into essentially the form in which it persists to this
day. With minor changes in terminology (Negrito is often substituted for
"Semang"; Senoi has replaced the pejorative "Sakai"; Proto-Malay, the too-
limited "Jakun"), this classification remains the unchallenged Malayan trinity
employed in ethnological descriptions of the aboriginal groups of the peninsula
(e.g., Carey 1976; LeBar, Hickey, and Musgrave 1964).

Schebesta ascribed the following racial, linguistic, and cultural attributes to each of the three ethnic groups:

Semang (Negritos): Very short (under 1500 mm) with long trunk and arms; dark, almost black skin color; dark-colored, wooly or frizzy hair. Now speakers of distinctive Austroasiatic languages having somehow lost their original Negrito language. Culturally very primitive nomadic hunters-and-gatherers who have retained their original monotheistic religion.

Senoi (Sakai): Slighter in build but taller and with fairer skin than the Semang. Wavy, dark-colored hair displaying a chestnut tint. Speakers of Austroasiatic languages closely related to those spoken in Indochina. Culturally more evolved than the Semang, the Senoi are shifting cultivators living in large communal longhouses and believing in animistic religions.

Proto-Malays (Jakun): Closely resemble the Muslim Malays with heavier body build and darker skin color than the Senoi. Straight, dark-colored hair. Speakers of Austronesian languages, the Proto-Malays are culturally the most evolved aboriginal group, engaging in large-scale horticulture with a complex social organization. They are animists.

This classificatory scheme reflects the theoretical assumptions of the Vienna school of anthropology, in which ethnic groups are defined by a set of fixed biological, linguistic, and cultural criteria. Moreover, each group is assigned its proper place in a cultural-historical sequence reflecting unilinear evolutionary assumptions. Schebesta assumed not only that social, linguistic, and cultural traits comprise stable associations that persist through time and across space, but also that the aboriginal ethnic assemblages he identified in Malaya represent, from the cultural-historical point of view, "three stages of development, although it must not be assumed that any one of them is based upon another" (Schebesta 1926:276). In his scheme, the supposedly primitive Semang are seen as the earliest wave of migrants to reach Malaya.[1] Subsequently the more advanced Senoi arrived and pushed the Semang into marginal areas of the peninsula only to be displaced in turn toward the margins by the Proto-Malays.

That this classificatory scheme should still be used more than half a century after it was formulated is little short of amazing, suggesting that Southeast Asia may be a marginal area offering refuge not just to primitive ethnic strata but to discredited intellectual theories as well. Even at the time Schebesta was formulating his classification scheme, Boas and other American anthropologists had already demonstrated that racial, linguistic, and cultural traits vary independently rather than constitute stable genetic associations. Today it is also recognized that most of these traits are subject to natural selection and undergo continual change in response to environmental forces. Despite these rather fundamental changes in our theoretical assumptions about the nature of ethnic groups, we go right on using Schebesta's classification scheme in our discussions of Malayan ethnology.

One can argue, of course, that the theoretical foundations of Schebesta's classification system do not matter as long as it works empirically in bringing order to what otherwise would be an intellectually unmanageable chaos of hundreds of different bands and tribal groups. Our minds can cope with a three-part but not a three-hundred-part ethnic universe, which is what Malayan ethnology consisted of before Skeat and Blagden and Schebesta introduced their tripartite classification. Moreover, this scheme does seem to work in practice, at least much of the time. One can go into the rain forest in the northern part of the peninsula and actually meet short, dark-skinned, frizzy-haired people who hunt and gather and appear to have a very primitive way of life—genuine Semang, as it were. One can then climb the slopes of the Main Range and see tall, fair-skinned, wavy-haired folk clearing swidden fields and living in longhouses—real-life Senoi. And in the south one can visit large horticultural settlements inhabited by Malay-speaking animists with straight hair—authentic Proto-Malays.

When I first started fieldwork in 1975 among the aborigines in Malaysia, I accepted Schebesta's categories as valid ones and was generally able to correctly identify the groups with which I worked. During visits to dozens of different aboriginal groups, I slowly came to question the validity of the Malayan trinity. What disturbed me initially was the extent of the physical diversity that I encountered. True, I actually did see short, dark-skinned, frizzy-haired Semang, but I also began to notice tall, fair-skinned, wavy-haired Semang, and short, dark-skinned, wavy-haired Semang, and every other possible combination. Meanwhile, on visits to groups labeled as Senoi, I encountered not only tall, fair-skinned, wavy-haired Senoi, but also short, dark-skinned, frizzy-haired ones, and again, all sorts of other combinations. Clearly, such observations did not fully coincide with prevailing theory; in such cases, the usual appropriate scientific strategy is to rethink one's theoretical concepts.

In fairness to earlier ethnologists, I must admit that I was not the first to notice this lack of fit between theoretical categories and real-world variability. Skeat and Schebesta had recognized considerable physical variability within groups. They attributed this variability to the recent mixing or hybridization of formerly pure groups, the standard explanation used by ethnologists of their time to account for diversity.

A. Terry Rambo

An Alternative Hypothesis for Malayan
Aboriginal Ethnogenesis

An alternative hypothesis can satisfactorily account for the existence of such intragroup variability. This hypothesis, which may be referred to as the "adaptive radiation model," can also explain the distributional patterns and many of the observed physical and cultural characteristics of aboriginal ethnic groups in Malaya. According to this hypothesis, present-day intra-group diversity reflects the incomplete adaptation of these groups to specific local environments following their partial isolation from a common aboriginal gene and cultural pool some thousands of years ago.[2] If this hypothesis is correct, then the diversity that we observe today is not the result of recent intermixture of previously pure migratory waves but instead reflects the incomplete evolutionary formation of several distinctive new ethnic groups out of a single ancestral population.[3] We are, so to speak, observing "species" in formation rather than disintegration.

Having proposed *in situ* differentiation resulting from adaptive radiation as an alternative hypothesis to the traditional migratory waves model of aboriginal ethnogenesis, let me hasten to point out that reality is probably not that simple. I believe that the modern ethnographer actually confronts the results of both processes (migration and hybridization on the one hand, and evolutionary divergence on the other) working simultaneously. Thus, while the Semang and the Senoi could have evolved out of a common ancestral population, the Proto-Malays are probably relatively recent immigrants from Indonesia, who, after arriving in the peninsula, exchanged genes and cultural traits with the earlier inhabitants (Figure 2.1).

The Dynamics of Ethnic Group Differentiation in Malaya

The hypothesis that the Semang and the Senoi have developed from a common ancestral population is based on the argument that while prior to the agricultural revolution they occupied a common ecological niche, they now occupy radically different niches. I am here using "niche" in the sense proposed by Hutchinson (1965) of an n-dimensional hypervolume whose parameters reflect the totality of environmental influences affecting survival of a species, rather than in the more common sense of the functional role or occupation of the species in the community.

Before the appearance of agriculture in the Malay peninsula several thousand years ago, the aboriginal population was probably largely confined to coastal and riverine zones with the vast interior forests essentially uninhabited. This forest is, from the standpoint of terrestrial mammals, virtually a biological desert. Carbohydrate and protein supplies are scant, widely scattered, and temporally unreliable. Even skilled hunters and gatherers find the interior forest to be an inhospitable habitat.[4] As Schebesta (1926:276) states, the Semang

> . . . live in a kind of symbiosis with [agricultural Malay and Senoi] villages and settlements, for, in my opinion they are unable to subsist solely on the roots they find in the jungle.

The development of agriculture allowed significant human occupation of the interior forests for the first time. The yams, bananas, rice, and millet raised in swidden plots provided the carbohydrates that were in short supply in the natural ecosystem. At the same time, shifting cultivation began to convert the previously undifferentiated forest into a mosaic environment having far greater capability than primary forest to support game animals, particularly pig, deer, and jungle fowl, thus greatly enhancing protein availability. Even today, the swidden-farming Senoi are generally much more successful in hunting than are the "hunting-and-gathering" Semang.

The swidden farming populations that began to settle the interior mountain ranges moved into a niche very different from that of their lowland ancestors. Under the new selective pressures encountered in the interior, the Senoi began to emerge as a biologically and culturally distinctive group. Later, the lowland coastal areas of the peninsula were settled by immigrant agricultural populations we now identify as Malays, who developed strong overseas trade relationships with Indonesia, India, and China. Products of the interior rain forest such as peafowl feathers, bee's wax, various resins, and rhinoceros horns became highly valued export commodities that were exchanged for manufactured goods such as porcelain and metal tools. This overseas trade created a new niche for specialized collectors in the lowland rain forests (Dunn 1975), a niche that I suspect was filled by aborigines displaced from the coastal zones by the agricultural immigrants. It is from this lowland forest population who lived by collecting wild products to trade for tools and food that the modern Semang evolved.[5]

So far I have been dealing in what, given the almost total lack of reliable archaeological evidence, can most charitably be called "conjectural history," a hypothetical reconstruction that is neither less nor more convincing than that proposed by Schebesta. I would now like to go beyond this kind of argument and show why, on ecological grounds, the hypothesis of local evolution of racial and cultural diversity is a plausible one. This is because the biophysical and cultural differences observed between the Semang and the Senoi are precisely those that would be expected to occur given the characteristics of the niches they occupy.

Ecological Niches and Ethnic Differentiation

If a map showing the territories occupied by aboriginal groups is overlaid on a topographical map, an extremely interesting correlation emerges: much of the area occupied by Senoi is over 500 meters in elevation (see Map 2.1). In contrast, the Semang are almost entirely found at elevations below 500 meters.[6] This means, among other things, that the two groups are exposed to radically different ambient temperature regimes. The annual mean temperature at Cameron Highlands in the heart of the Main Range is 64°F compared to 78°F at Kuala Lipis in the lowlands. Perhaps more important are the differences in daily and annual range. The uplands range from 55°F to 73°F, with the absolute low being 36°F; and the lowlands range from 71°F to 90°F, with the absolute low being 63°F (Dale 1974).

A second major difference in the two niches is that the Semang spend almost all of their time living under the dense forest canopy, whereas the Senoi live in cleared settlements and spend much of their time working in relatively open swidden fields. This means that the Senoi live under conditions of great diurnal temperature variation, low relative humidity (55 percent at midday), and high solar insolation, whereas the Semang live under an extremely stable temperature regime, with constantly high relative humidity (never dropping below 75 percent), and virtually no direct exposure to solar radiation (Rambo 1985 presents comparative temperature and humidity data for a swidden plot and the deep forest).

The nutritional parameters of the two niches also differ significantly. The Senoi have reliable access to adequate supplies of calories and protein. The Semang, in contrast, face an unreliable and sometimes insufficient supply of both calories and protein.

The two niches also differ in their requirements for human muscular activity: The primary demand on the Semang is for high mobility, especially the ability to move rapidly through the forest undergrowth. Their resource collection activities involve using the blowpipe, climbing trees to obtain fruit, and digging wild tubers, with only the latter activity calling for any great muscular power. Tree climbing is not without danger, and smaller individuals are less likely to suffer serious injuries from falls than are larger ones. The Senoi, on the other hand, face extremely heavy work demands for clearing swidden plots, planting, and performing other muscle-powered agricultural activities.

A final difference between the two niches until the establishment of British colonial authority lay in the extent and intensity of human predation pressure. The Semang were frequently hunted as slaves by the coastal Malays and by the Thais in the north, whereas those Senoi, such as the Temiar, who resided in the extremely isolated and rugged mountains of the Main Range enjoyed relative immunity from slave raiding (Endicott 1983).[7]

If we compare these ecological factors to the physical characteristics of Schebesta's ideal type Semang and Senoi, we find that all differences between the two groups are in the directions predicted by ecological principles (Table 2.1). The physical characteristics of the Semang represent adaptations to the niche occupied by these nomadic foragers in the hot, humid, dimly lit lowland tropical rain forest.[8] The Senoi appear physically well adapted to the niche occupied by farmers in cooler, less humid, sun-lit fields in the mountains. Admittedly, correlation does not establish causality, but in this case I doubt that we are dealing with the accidental outcome of random events.

There are similar patterns of correlation between the cultural characteristics of the two groups and their respective niches (see Benjamin 1985). Thus, the Semang appear to be very primitive, not because they represent a surviving Paleolithic stratum that has been pushed into an isolated marginal refuge area, but rather because a nomadic foraging adaptation is both the most profitable and the safest strategy for a defensively weak minority ethnic group living close to militarily dominant, and often hostile, agriculturalists. The adaptation is profitable because under the terms of trade the Semang can obtain more food calories for less labor expenditure by exchanging forest products for their Malay neighbors' surplus rice than they can by growing the rice themselves. From the standpoint of security, the adaptation also makes sense because nomads are much harder to catch than settled farmers (trading may have been carried out in precolonial times by means of the so-called "silent barter" in which the trading partners never came into direct contact).

The nomadic foraging adaptation of the Semang imposes a whole set of functionally related cultural requirements that we normally associate with a primitive hunting-and-gathering evolutionary stage (e.g., small population size, limited cultural inventory, simple and small-scale social and political organiza- tion). In the same way, the swidden farming-based adaptation of the Senoi has its own functionally derived cultural correlates and the horticultural adaptation of the Proto-Malays yet another set. Such configurations are not directly caused by the physical environment in the sense that the determinist school used to claim, but they can only evolve and persist within a certain limited range of ecological conditions.

Place any aboriginal population into the kind of niche filled by the Semang and sooner or later they will, in effect, become Semang. Conversely, put a group of Semang into the Senoi niche, and they will change in the direction of becoming Senoi. As is amply documented by several other chapters in this volume (e.g., Cadeliña, Foster, Yengoyan), adaptive change of this sort in cultural patterns is commonplace not just in Malaysia but throughout Southeast Asia. Seemingly more problematic is the possibility of rapid change in biophysical characteristics of the sort required for differentiation of the Semang and the Senoi in the relatively short period since the introduction of agriculture and the forest product trade created their distinctive niches.

Livingstone (1958) has made a persuasive case for the spread of the sickle-cell gene in tropical African populations within only a few thousand years since their adoption of swidden agriculture and consequent increased exposure to the selective pressure of endemic malaria. Much less is known about the rate of change for genetically more complex physical characteristics such as body size and proportions, hair form, and skin color. Evidence from other animal species suggests, however, that such change can be rapid, even under what would seem to be rather mild selective pressure. European house sparrows, for example, have developed local ecological races within 100 to 200 generations of their introduction into North America (Johnstone and Selander 1964). This would be roughly the number of human generations since the development of agriculture and overseas trade in forest products opened distinctive niches for the Semang and the Senoi in Peninsular Malaysia. More relevant to the case of the Malaysian aborigines, but less well documented than the case of the house sparrows, is the physical differentiation that has evidently occurred in American Indian populations during the approximately 15,000 years since their ancestors first crossed the Bering Straits from Asia. Newman (1953) claims there are definite clines in body size and proportions from higher latitude populations to those dwelling in the tropics, which appear to reflect the outcome of climatic selection according to Bergmann's and Allen's Rules. Even more relevant to this case is Milton's report (1983) that Maku forest dwelling hunter-gatherers in northwestern Amazonia are both shorter and lighter than neighboring Tukanoan swidden farmers. If Milton is correct in attributing the small body physique of the Maku to selection by factors associated with life in the equatorial forest, this differentiation must have occurred in the few thousand years since agriculture developed in Amazonia, a situation that appears exactly parallel to that of the Semang and Senoi in Malaya.

I am under no illusion that the arguments presented here have in any sense proved the case for local differentiation of the Semang and the Senoi into distinct ethnic groups in response to differing environmental selective pressures. The available empirical information is simply inadequate for this task. What I do hope to have shown, however, is that a hypothesis of local evolution is at least as plausible as the conventional hypothesis that explains Malayan aboriginal ethnic differentiation in relation to migratory waves. This brings us back to the question posed in the title of this paper, "Why are the Semang?" By now it should be clear that my answer to that question is that they are Semang because they are there and not, as Schebesta argued, that they are there because they are Semang.

Acknowledgments

An earlier version of this paper was presented to a departmental seminar of the University of Hawaii Department of Anthropology in 1980. The field research on which it is based was supported by a series of University of Malaya Staff Research Grants. The continuing support provided this work by Prof. Yip Yat-Hoong, the former Deputy Vice Chancellor for Research of the University of Malaya, is gratefully acknowledged as is the fine cooperation provided by the staff of the Department of Orang Asli Affairs, especially its Director General, Dr. Baharon Azhar bin Raffie'i. Much valuable research assistance was provided by my students, particularly Alberto G. Gomes and Koh Bee Hong, at the University of Malaya Department of Anthropology and Sociology. Kirk Endicott and George W. Lovelace made useful comments on earlier drafts of this paper. Continuing discussions with Karl L. Hutterer have greatly influenced my thinking about the ecology of Southeast Asian foraging societies. The possible adaptive value of smaller body size to the Semang in reducing injuries suffered as the result of falling out of trees was called to my attention by Frank B. Livingstone.

Notes

1. The belief that the Semang are survivors of the most ancient migratory wave to enter Malaya is seemingly supported by the existence of "Negritos" elsewhere in Southeast Asia. These include the Aeta or Agta of the Philippines, which are discussed by Eder and by Cadeliña in their chapters in this volume, and the aboriginal inhabitants of the Andaman Islands. When Schebesta began his research in Malaya, he held to Schmidt's belief that these groups were part of a worldwide "pygmy" wave of migration derived from Africa. He subsequently rejected this view in favor of seeing the Southeast Asian Negritos as an independent racial-cultural group having no connection to the African pygmies. Discussion of the "pan-negrito" hypothesis is beyond the scope of this chapter. I would simply note that the evidence supporting a common origin for all Southeast Asian Negritos is tenuous at best. No detailed comparative seriological studies have been done so that the argument for common origin is based on sharing of visible physical characteristics (short stature, dark skin, and frizzy hair form) and cultural traits (e.g., bow and arrows, beehive huts, belief in a thunder god) (Cooper 1940). Many of these characteristics can be shown to be subject to environmental selection, making it hardly surprising that they are shared by tropical forest foragers. Others are shared with neighboring non-Negrito groups and appear to be derived from a common Southeast Asian

cultural background. The curious fact that the different Negrito groups speak languages closely related to those of their non-Negrito neighbors, rather than sharing a common language of their own, also raises questions about the assumption of common origins. Also puzzling—if one assumes common ancestry for the Andamanese, Aeta, and Semang—is the total absence of any traces of Negritos in the vast areas of tropical forest lands, notably on the islands of Borneo and Sumatra, which separate these groups.

2. Geoffrey Benjamin (1985:233) has also suggested that the three distinctive "modes of production" associated with the Semang, Senoi, and Malay "probably differentiated out of an earlier ('Hoabinhian') sociocultural matrix." Benjamin's emphasis is on cultural differentiation, rather than racial differentiation which is my main concern in this chapter, but we share the underlying assumption that differentiation has occurred in place rather than from migrations.

3. The racial identification of this ancestral "proto-Orang Asli" population is an open question in the absence of a much better archaeological record than we now possess. No prehistoric Negrito skeletons have been discovered in Malaya nor is there any physical evidence for the presence of "Australoids" or "Austromelanesians," asserted by physical anthropologists in the past. Indeed, Bulbeck (1981) makes a convincing case for including the prehistoric inhabitants of the Malay peninsula within a general Southeast Asian population mosaic out of which all modern groups in the region have evolved. His analysis of skeletal material from the Hoabinhian site of Gua Cha in the highlands of northern Malaya indicates an average stature somewhat greater than that of modern Semang and Senoi. Skin color is obviously impossible to reconstruct from skeletons, but it is not unreasonable to assume that the average shade of lowland Southeast Asians was relatively dark before the massive inflow of Chinese genes in recent centuries. Hair form is also unknown, but as Bellwood (1979:46) suggests, frizzy hair may have been more widely distributed in Southeast Asia before the inflow of northern Mongoloid genes. The "proto-Orang Asli" population would thus appear to have contained sufficient genetic variability to have allowed the rapid evolution of its modern components once these became ecologically specialized.

4. Kirk Endicott, an ethnographer who has conducted extended field research on the Batek Semang, states (pers. com.) that he ". . . is not convinced that it was virtually impossible for people to live in the interior (low-land) forests by foraging alone. . . . For one thing, there are at least ten species of Dioscorea eaten by the Batek, three of which are quite abundant in the lowland forests." In addition, ". . . there are lots of fish, frogs, and crustaceans in the streams and quite a few monkeys in the certainly enough

food there for at least a few small nomadic groups to survive indefinitely, though it is far more efficient energetically to trade for some foods in some circumstances. . . ." He further notes that, ". .the Batek have lived quite happily for long periods with minimal contact with farmers or outside food sources, even in the recent past" and adds that he ". . . almost never heard nomadic Batek complain about hunger unless a lot of people were sick." In the absence of systematic empirical assessment of the long-term human carrying capacity of the Malayan rain forest, it is impossible to resolve the question of whether or not foraging populations can persist indefinitely without any trade relations with agriculturalists. It may be that well-endowed areas can sustain small groups, such as the Batek observed by Endicott, for relatively short periods but become depleted if intensively exploited for many years, hence the pulse-like nature of Semang trade relations with their settled neighbors.

5. Geoffrey Benjamin's (1976) reconstruction of the historical development of the Semang and Senoi languages offers interesting linguistic evidence in support of this hypothesis. He argues that the principal Semang language group ("Northern Aslian") and the Senoi language group ("Central Aslian") have evolved from a common ancestral proto-Aslian language. Using lexico-statistics, he dates the beginning of their divergence at between 6610 and 6410 B.P., a period that coincides with the introduction of agriculture.

6. The distribution of these aboriginal groups is derived from Benjamin's (1985) map of the approximate distribution of Orang Asli groups. These indicate cultural areas rather than precise location of settlement or populations. In Map 2.1, the distribution of Senoi is based on Central Aslian, excluding Lanoh; the distribution of Semang is based on Northern Aslian and Lanoh.

7. Unfortunately for the neatness of my hypothesis, Endicott also states that the Semai-Senoi who lived in the foothills on the western slopes of the Main Range were exposed to the most intense slave raiding of any aboriginal group.

8. In a Master's thesis, which came to my attention after I presented this paper at the conference, Bulbeck (1981) suggests that a number of "Negrito" physical characteristics represent rain-forest adaptations rather than indicators of racial phylogeny. Thus, dark skin color is inconspicuous in a dim environment and short stature favors easy movement. He also suggests that frizzy hair helps to keep the head warm and dry in a very moist environment (Bulbeck 1981:418–421), whereas I have suggested that it is a cooling device under hot, humid atmospheric conditions. It should be relatively simple to do the empirical testing necessary to establish which of our hypotheses, if either, is correct. I am grateful to Kirk Endicott for supplying me with a copy of Bulbeck's thesis.

References

Bellwood, P.
    1979 *Man's Conquest of the Pacific: The Prehistory of Southeast Asia and
         Oceania.* New York: Oxford University Press.

Benjamin, G.
    1976 Austroasiatic subgroupings and prehistory in the Malay peninsula.
         In *Austroasiatic Studies*, edited by P.N. Jenner, L.C. Thompson,
         and S. Starosta. Honolulu: University Press of Hawaii. Pp. 37–128.
    1985 In the long term: Three themes in Malayan cultural ecology. In
         Cultural Values and Human Ecology in Southeast Asia, edited by
         K.L. Hutterer, A.T. Rambo, and G. Lovelace. Ann Arbor: University
         of Michigan Center for South and Southeast Asian Studies, Paper
         No. 27. Pp. 219–278.

Bulbeck, F.D.
    1981 Continuities in Southeast Asian evolution since the Late Pleis-
         tocene. M.A. thesis, Australian National University, Canberra.

Carey, I.
    1976 *Orang Asli: The Aboriginal Tribes of Peninsular Malaysia.* Kuala
         Lumpur: Oxford University Press.

Cooper, J.M.
    1940 Andamanese-Semang-Eta cultural relations. *Primitive Man* 13:29–
         47.

Dale, W.L.
    1974 Surface temperatures in Malaya. In *Readings on the Climate of
         West Malaysia and Singapore*, edited by Ooi Jin Bee and Chia Lin
         Sien. Singapore: Oxford University Press. Pp. 57–73.

Dunn, F.L.
    1975 *Rain-Forest Collectors and Traders: A Study of Resource Utilization
         in Modern and Ancient Malaya.* Kuala Lumpur: Malaysian Branch
         of the Royal Asiatic Society, Monograph No. 5.

Endicott, K.
    1983 The effects of slave raiding on the aborigines of the Malay peninsula.
         In *Slavery, Bondage and Dependency in Southeast Asia*, edited by
         A. Reid and J. Brewster. Brisbane: University of Queensland Press.
         Pp. 216–245.

Hutchinson, G.E.
 1965 *The Ecological Theater and the Evolutionary Play.* New Haven: Yale University Press.

Jimin bin Idris
 1968 Distribution of Orang Asli in West Malaysia. *Federation Museums Journal* 13:44.

Johnstone, R.F., and R.K. Selander
 1964 House sparrows: Rapid evolution of races in North America. *Science* 144:548-550.

LeBar, F.M., G.C. Hickey, and J.K. Musgrave
 1964 *Ethnic Groups of Mainland Southeast Asia.* New Haven, CT: Human Relations Area Files Press.

Livingstone, F.B.
 1958 Anthropological implications of sickle cell gene distribution in West Africa. *American Anthropologist* 60:533-562.

Milton, K.
 1983 Morphometric features as tribal predictors in North-western Amazonia. *Annals of Human Biology* 10:435-440.

Newman, M.T.
 1953 The application of ecological rules to the racial anthropology of the aboriginal New World. *American Anthropologist* 55:311-327.

Rambo, A.T.
 1985 *Primitive Polluters: Semang Impact on the Malaysian Tropical Rain Forest Ecosystem.* Museum of Anthropology, Anthropological Paper No. 76. Ann Arbor, MI: University of Michigan.

Schebesta, P.
 1926 The jungle tribes of the Malay peninsula. *Bulletin of the School for Oriental and African Studies* (London) 4:269-278.

Skeat, W.W., and C.O. Blagden
 1906 *Pagan Races of the Malay Peninsula.* London: Macmillan.

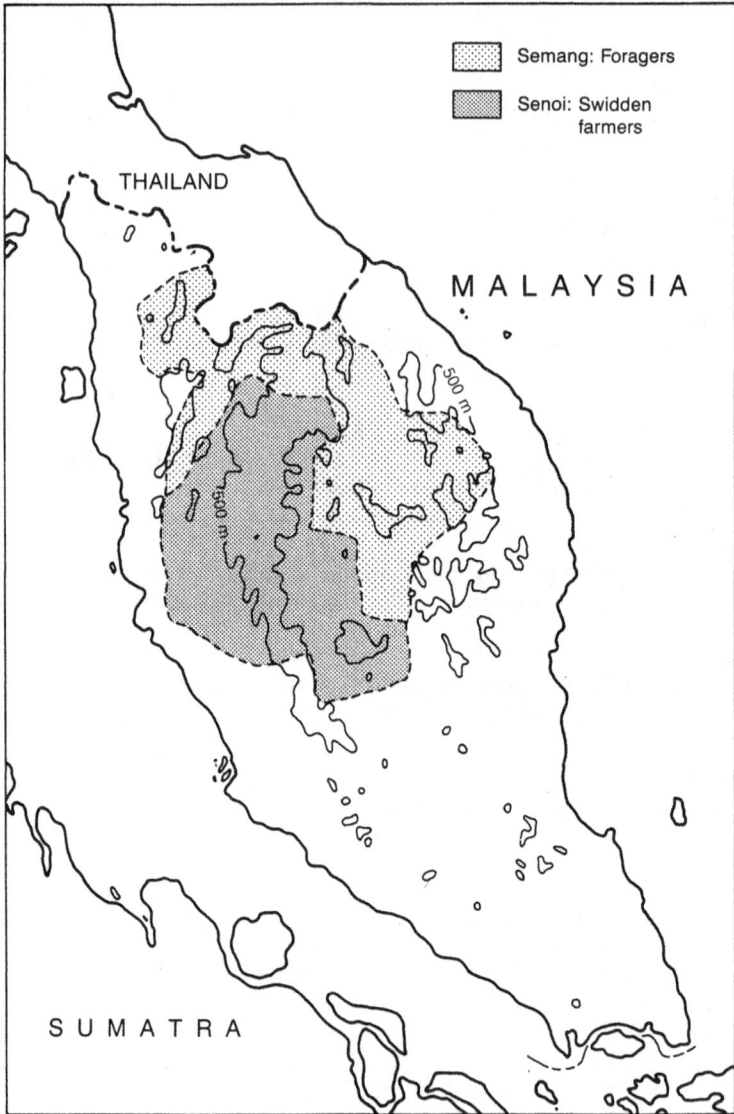

Map 2.1
Distribution of Malayan aborigines
in relation to highland areas

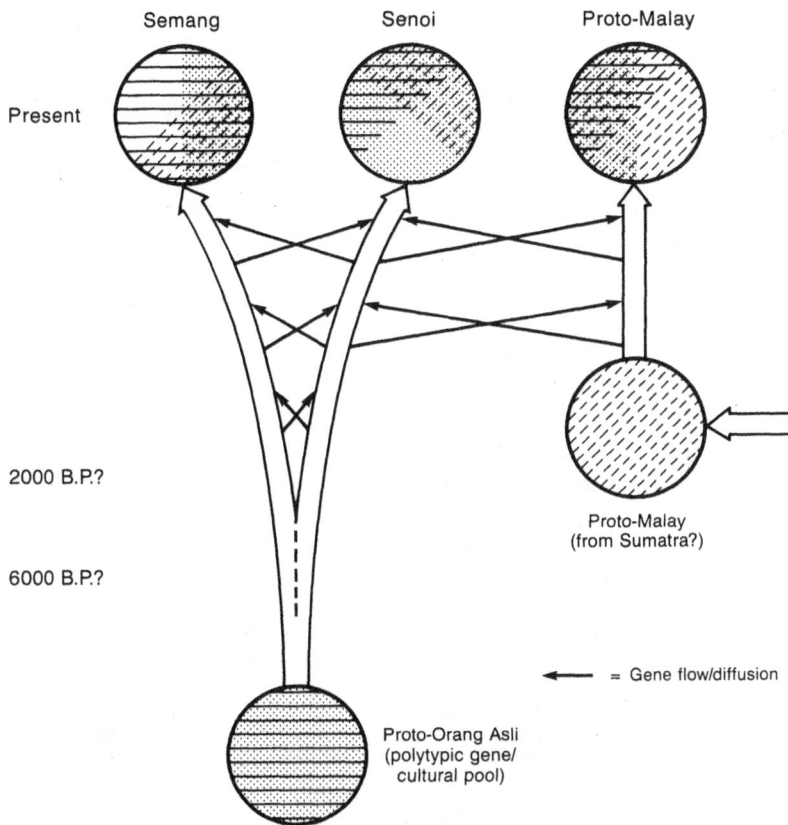

Figure 2.1
Synthetic model of Malayan aboriginal ethnogenesis

Table 2.1
Comparison of ecological niches and physical
adaptations of Semang and Senoi

Factor	Semang	Senoi
Ambient temperature	Constant, warm: average 78°F, daily range 71–90°	Variable, cool: average 64°F, daily range 55–73°
	Small stature and elongated extremities increase rate of metabolic heat loss (Bergmann's and Allen's Rules)	Larger stature reduces metabolic heating requirements
	Dark skin color may enhance radiative heat loss at certain wavelengths	
	Frizzy hair form permits increased cranial heat radiation	
Relative humidity	Constant, high: average 85–90%	Variable, low: 55–75%
	Flattened nose with broad nostrils (nasal index 97)	Longer narrower nostrils (nasal index 89) moisten drier air
	Dark skin color (Gloger's Rule)	
Solar radiation	No direct exposure under forest canopy	Extended exposure to intense radiation in high-elevation swidden clearings
		Light skin color and chestnut hair color increase reflectivity and reduce heat load

Factor	Semang	Senoi
		Wavy hair form shields head from direct radiation
Nutritional stress	Supply of carbohydrates limited and erratic	Supply of carbohydrates adequate and reliable
	Small stature minimizes caloric demands	
Mobility	Nomadic foraging requires frequent, rapid, cross-country travel through forest undergrowth	Sedentary agriculture requires only limited cross-country travel
	Small stature allows easier movement through forest undergrowth and reduces likelihood of injury from falling out of trees	
Labor power	Foraging and blow-pipe hunting requires light labor input	Swidden farming demands heavy labor input
		Larger stature gives increased leverage in using axes and dibble sticks
Human predation	Slave raiding by neighboring Malays a constant threat (until late 19th century)	Slave raiding infrequent in isolated mountain settlements
	Dark skin color is effective camouflage under forest conditions	Camouflage not required

CHAPTER 3

HUNTER-GATHERER/FARMER EXCHANGE IN THE PHILIPPINES: SOME IMPLICATIONS FOR ETHNIC IDENTITY AND ADAPTIVE WELL-BEING

James F. Eder

An often-remarked characteristic of prehistoric and contemporary Southeast Asian hunting-gathering populations concerns the nature and degree of their economic ties with surrounding agricultural populations. Especially in the Philippines, exchanges between Negritos and lowland Filipinos of forest products such as wild meat, honey, Manila copal, and rattan for agricultural foods and trade goods have been widely reported and indicate a considerable degree of economic interdependence.

Recently, Peterson (1978) has placed food exchanges between a particular Negrito group, the Agta of northeastern Luzon, and their sedentary agricultural neighbors, the Palanan, in an analytical framework that emphasizes not simply the *interdependent* but the *mutually beneficial* nature of that exchange. In particular, exchanges of Agta nondomestic protein foods (wild pig, deer) for Palanan domestic carbohydrate foods (corn, yam, manioc) allow each population to specialize, in effect, in what it does best. The Agta find it relatively easy to produce a surfeit of protein foods but suffer chronic shortfalls in carbohydrate foods, while the reverse is true of the Palanan. By matching their respective shortfalls and surfeits against those of their neighbors, such exchanges, it is argued, are energetically efficient and hence make "good sense" for the parties involved. Although Peterson is ambiguous about the degree to which she is generalizing, there is at least the implicit suggestion that her analysis might apply to other cases of hunter-gatherer/farmer exchange as well.

Peterson's (1978) paper is valuable for calling attention to—and quantifying—the considerable degree of economic interdependence between the members of the two societies in question. But her analysis is cast in a

37

functionalist mode that, in my opinion, commits two classic functionalist errors. First, it emphasizes benefits and ignores costs; second (despite a surface appeal to evolutionary theory), it is fundamentally ahistorical and nonevolutionary. My purpose in this chapter is to discuss these two shortcomings and hence to place the analysis of hunter-gatherer/farmer exchange in the Philippines on a more realistic footing. With respect to the first shortcoming, I want to discuss some of the economic, social, and cultural *costs* to Philippine Negrito populations of entering into trade relationships with lowland Filipinos. In this discussion I will draw heavily on my own research among the Batak, a Negrito group inhabiting the mountains of northern Palawan Island. I too have emphasized the extent to which the Batak participate in a wider socioeconomic system, but my analysis focuses on the manner in which such "participation"—i.e., the entire web of Batak-outsider relationships, including trade in food—has severely stressed and disrupted all aspects of Batak adaptation (Eder 1977a, 1977b, 1978, 1984, 1987). I will also draw on my own observations and the reports of others concerning comparable difficulties among other Philippine Negritos. It is not, however, my intention to enter an argument about whether the "costs" of exchange relationships outweigh the "benefits" to Negrito populations, but only to emphasize that significant costs do exist.

With respect to the second shortcoming, however—the ahistorical and nonevolutionary nature of Peterson's (1978) analysis—I will be more contentious. I will argue that Peterson's analysis, by reifying cultural boundaries and subsistence economic types, by emphasizing complementarity and voluntarism, and by being frozen at a particular point in time, overlooks a vital, broader pattern of historical, political economic changes involving Agta and Palanan—and Negritos and lowland Filipinos generally. Consideration of this broader pattern, I argue, suggests that what appear to be symmetrical, complementary "exchange" relationships between two otherwise independent "populations" have the potential to become classlike ethnic relations within a single society, with attendant marginalization of the "hunter-gatherers" in question. They, in turn, to the degree that they persist at all as a discrete ethnic group, find their ethnicity to be redefined—less and less in terms of linguistic distinctiveness or some other positive cultural boundary marker, and more and more in terms of their reduced socioeconomic circumstances and their very activities, as producers for exchange, of forest products needed by a wider socioeconomic system. In short, Peterson fails to recognize (or at least to discuss) the potential for seemingly symmetrical "exchange relationships" to evolve over time into something quite different—i.e., into something that threatens the ethnic identity, the adaptive well-being, and even the actual survival of the Negrito populations involved.

I will begin with a brief summary of Peterson's analysis. Then I will lay out the details of my own arguments just discussed. In the final section of the chapter, I will bring my approach to bear on the related question of why Philippine Negrito populations, while all practicing at least a modicum of agriculture, have generally failed to "evolve" into full-time farmers.

Peterson's Analysis

Peterson's (1978) paper is based on her own field observations of Agta-Palanan interactions in northeastern Luzon, a reading of the Mbuti-Bantu relationship as described by Turnbull (1965), and a survey of the literature on hunter-gatherer/farmer exchange, particularly in the Philippines. My primary concern here is with her analysis and use of the Agta-Palanan case.

Peterson describes the Palanan Bay watershed, the site of her fieldwork, as "one of the more isolated and environmentally least hospitable areas of the Philippines" (1978:337). Here are found "two ethnically and physically distinct populations": about 800 Agta, Negrito hunter-gatherers known to others as Dumagat or Aeta; and about 10,000 Palanan, peasant farmers apparently long resident in the area but who since World War II have expanded inland and encroached upon traditional Agta territory (1978:342). Neither of these populations is self-sufficient in food production. Palanan farmers produce an abundance of carbohydrate foods (mostly corn, yam, and manioc) but either do not or cannot raise enough livestock to supply their animal protein needs (1978:338–339). The Agta, in contrast, obtain meat and fish in abundance but find it difficult to obtain starch staples. This situation presents, in Peterson's words, "optimum opportunity for economic interdependence; the Agta are a people who produce limited carbohydrate or other vegetable foods, and the Palanans produce limited protein foods" (1978:339).

Peterson portrays the resulting exchange relationship as voluntary and based on a simple cost-benefit calculus by each party, rather than on coercion or any obsessive Agta preference for agricultural foodstuffs (1978:337). Her own summary of the "importance of intercultural exchange to the Agta" covers the following points:

- Protein surfeit represents at least as great a problem as protein deficit.
- Obtaining adequate carbohydrate foods is a significant problem.
- Exchange allows the continued maintenance of the low labor input hunting-gathering lifeway.
- Any Agta emphasis on farming would not be compatible with their preexisting emphasis on hunting-gathering (due to farming's characteristic labor demands and scheduling requirements).
- Any Agta emphasis on farming would increase competition with Palanan for land.

- The existing multidimensional subsistence strategy (mixed hunting-gathering, limited farming, and exchange) represents "an effective adaptation for a population with limited technology" (Peterson 1978:343–344).

Peterson never addresses the question of possible game depletion by Agta specializing in the production of wild meat for exchange. She does briefly suggest that "population interdependence" might have implications for the notion of carrying capacity, but after adducing no relevant data on the subject she curiously concludes that "Agta-Palanan relations certainly appear to be 'ecologically sound'" (1978:346). Peterson also explains that there are other dimensions to the Agta-Palanan relationship than food exchange. Palanan regularly hire Agta to perform farming chores (1978:342), but although Peterson notes that other observers have called attention to the potential for abuse in such relationships (1978:336), she makes no mention of such problems in the Agta-Palanan case. Finally, she does acknowledge the presence of a certain amount of conflict between Agta and Palanan over land, arising from the latter's interest in permanently acquiring and improving agricultural land, rather than merely using it for a spell for swidden-making, as the Agta do. Thus we are told that Palanans acquire the fallow swidden land of Agta "by two means: purchase and land grabbing" (1978:340), and we are told that "the politically and legally more sophisticated peasants encounter few obstacles when they choose to usurp Agta land without compensation" (1978:342).

But Peterson's overwhelming emphasis is on the *mutually beneficial* nature of the *exchange* relationships between Agta and Palanan. According to her, "Palanan-Agta exchange represents a labor specialization that coordinates the two populations in a higher order economic system. The efficacy of this specialization can be illustrated by exploring the integration of their relative economic roles" (1978:343). Again, "in short ... the union of culture-specific specialization and inter-cultural exchange allows 67 particularly effective utilization of land and labor resources" (1978:344).

A Contrasting Perspective

Peterson's interesting analysis is valuable, as I have said, for calling attention to the *importance* of hunter-gatherer/farmer exchange in the Philippines. But her single-minded emphasis on the mutual benefits entailed by such exchange is at best naive and at worst wrongheaded.

To explain why, let me begin by setting my own perspective on such "exchange" in the context of two broad generalizations about the status of Philippine Negritos today. First, however salutary the present nature of Agta-

Palanan exchange relationships may be, hunting-gathering is not exactly the wave of the future for Philippine Negritos (or for anybody else, for that matter). Such relevant factors as increased population pressure from land-hungry lowland settlers, the activities of multinational corporations, and gross destruction of forest cover should scarcely need mentioning. Second, whatever the present physiological, psychological, and cultural well-being of individual Agta, many (perhaps even most) Philippine Negritos are in badly reduced socioeconomic circumstances and at least some are extensively deculturated. As my qualifying comments indicate, I am less interested at this point in determining whether Peterson got the Agta-Palanan case wrong (indeed, I have never visited the Palanan Bay area) than in guarding against the possibility that others might assume that a comparable interpretation generally applies to Negrito-lowlander relationships.

In short, and drawing heavily, as I said, on my own fieldwork among the Batak, I would like to look over the present circumstances of Philippine Negrito groups and ask what role "hunter-gatherer/farmer exchange" might have had in explaining how they got that way. Throughout, I hope to illustrate the importance of seeing such exchange not merely as present or absent but as something that may develop and intensify over time—with ever-changing consequences for those involved.

Negrito/Lowland Filipino Exchange

Some Economic, Social, and Cultural Costs

I have deliberately recast the notion of "hunter-gatherer/farmer exchange" as "Negrito/lowland Filipino exchange" to emphasize that (1) a *range of exchanges* occurs between members of such populations, not just exchanges of wild meat for domestic carbohydrates, and (2) there is a basic *asymmetry* in all such exchanges, obscured by Peterson's emphasis on "complementarity," that arises because the hunting-gathering populations in question are usually small, isolated, and unsophisticated, while the "farming" populations in question carry behind them all the "structural weight," so to speak, of the legal apparatus and superior cultural capital of lowland Filipino society. This may overstate the case somewhat, but I do it because it immediately raises the question of whether the exchanges in question might routinely occur to the advantage of Filipinos and to the disadvantage of Negritos.

"Exploitation" is a thorny issue, and I do not want to imply that the present reduced socioeconomic circumstances of many Negritos arose because they have been "exploited"—although many observers have commented on the poor economic treatment Negritos have received from outsiders. But I would like to

call attention to the following extended passage from a recent dissertation about
the Agta by Rai (1982), in which the author specifically addresses and refutes
Peterson's (1976:329) claim that Agta/non-Agta exchange is symmetrical or, if
anything, is biased in favor of the Agta.

> I suggest that the symbiosis is biased in favor of the outsiders.
> This bias arises because the external dependence of the Agta
> today is crucial tó their economic survival. If the neighboring
> agricultural populations were called upon to make minor adjust-
> ments in their animal protein acquisition (and their seasonal
> labor requirement), they could maintain an independent
> economic system without their trade relation with the Agta.
> The Agta, on the other hand, have become virtually dependent
> on the outside system. Given their shift away from the
> traditional mode of life, the Agta cannot remain economically
> self-sufficient. . . . It is thus an imperative for the Agta to main-
> tain the trade or other economic relationship with outsiders even
> at the cost of their own economic exploitation and subordina-
> tion.

> The economic exploitation of the Agta by outsiders takes many
> forms. In Agta trade of forest items to outsiders, the latter
> control the market. For example, in spite of the acute demand
> for wild animal protein in the adjoining agricultural settlements,
> the price of wild animal meat is at least thirty-three percent
> lower than for domestic meat. In other barter trade, Agta are
> even more ruthlessly exploited. Three to nine wild pigs
> (approximate value 600 to 2,000 pesos) must be paid for one
> secondhand transistor radio (approximate value 200 pesos).
> Five hundred pesos worth of tree resin is bartered against a
> secondhand tape recorder (approximately 150 pesos value).
> Some outsiders justify such behavior by saying that Agta trade
> goods are wild and so require no work in raising them. The local
> middle man, who handles the purchase of trade items like copal
> resin and shellfish, often fixes the price as low as twenty percent
> of the local market. For example, a forty kilogram sack of copal
> resin is bought by the middleman for fifteen pesos and then sold
> to the contractor for fifty pesos. Most Agta are unfamiliar with
> the units of exchange and get further cheated. In wage labor,
> Agta are often paid less than their agricultural counterparts.
> Whereas an Agta is hired for only thirty pesos for four days of
> portering, an outside plowman summoned to work in the Agta

field may demand as much as half of the production of the field. (Rai 1982:191–192)

Rai (1982) also informs us that Agta/non-Agta exchanges are not limited to food items, for the Agta:

> spend a considerable proportion of their income from trade and wage labor in buying consumer goods. These consumer goods are becoming both a need and a status symbol to the Agta. Agta today buy items like soap, spices, carbonated drinks, coffee, sugar, canned food, nail polish and face powder. Alcoholism is another trend that is rising among the Agta. ... Agta regularly trade a part of their game for commercial liquor and in the coastal areas, for locally brewed coconut or nipa beverages. The outsiders exploit these acquired habits of the Agta. Often unsolicited, the outsiders extend credit in the form of consumer goods or beverages to the Agta. Once caught in the downward spiral of indebtedness, Agta can be seriously abused. (Rai 1982:193)

In my opinion, comparable patterns obtain almost anywhere in the Philippines where Negritos trade with outsiders. The best that can probably be said is that continued mobility allows most Negritos to escape the worst forms of economic abuse by hiding in the forest or by moving to new locales (e.g., Rai 1982:193; Hart 1978 makes a similar observation about Pygmy trade with outsiders).

There is a sense in which too much should not be made of the presence of "exploitation." For example, it can be argued that if an item bought for 15 pesos from the Agta brings 50 pesos elsewhere, that's capitalism, and the Philippines is a capitalist country. Further, that Agta and other Negritos are often exploited does not itself necessarily undermine Peterson's (1978) analysis. But more serious is the fact that Negrito/lowlander trade entails, for Negritos, a variety of opportunity costs and time allocation conflicts that simply make it less remunerative and "efficient" than it appears at first glance. Analyses, like Peterson's, which only compare "production" times for the goods to be exchanged (e.g., search, pursuit, and kill times for wild game; in-field agricultural labor costs), ignore a variety of ancillary activities and considerations necessary to the *maintenance* of exchange relationships.

A striking aspect of Batak behavior is the pervasive influence that exchange relationships—and concern about maintaining those relationships—have on all aspects of Batak economic life. Thus, Batak take time out from whatever they are doing to talk with or entertain their trading partners, whenever the latter visit a Batak settlement. Scheduling of other Batak subsistence activities is often subordinated to meeting the needs and demands of traders—individuals who are also, not incidentally, often creditors. Batak travel frequently back and forth to the lowlands to deliver goods previously promised, to collect payment for goods previously delivered, to obtain an advance against goods yet to be delivered, to inquire about future work or exchange possibilities, and so on. Such travel, which involves trips of several hours to a full day's duration (since Batak, once in the lowlands, are often dilatory about returning), figures strikingly in Batak time allocation budgets. Although the actual "returns" to particular episodes of travel may be difficult to identify, such travel *is* "all part of a day's work," and any accounting of the comparative cost-effectiveness of meeting subsistence needs through exchange relationships must take account of it.

A final economic cost to Negrito populations of reliance upon exchange relationships with outside peoples is the resource depletion that appears to accompany such exchanges. Possible contributing factors to such depletion include the production-for-exchange efforts of the Negritos themselves and the direct and indirect effects of the activities of encroaching non-Negrito populations on forest resources. Peterson (1981), it should be noted, has argued against this interpretation. Consistent with her emphasis on the functional benefits of hunter-gatherer/farmer interaction, she argues that agricultural expansion, because of a characteristic "edge effect," allows traditional Agta game species to *thrive* (e.g., in the "ecotone" of scrubby regrowth characteristically found between the forest and new swidden fields). Thus the intrusion of farming people into territory once exclusively inhabited by hunter-gatherers may contribute to the *persistence* of a hunting-gathering lifeway.

Other observers of the Agta, however, have strongly disputed Peterson's (1981) hypothesis, and I find their own data and arguments to be more convincing. Rai observed that agricultural expansion had no economic benefits for the Agta he studied but merely displaced them from their land and in fact did deplete their game resources (Rai 1982:184–188). Headland, who has worked for many years with the Casiguran Agta, says that wild pig and deer, the principal Agta game species, do not in fact thrive at the "edges" of lowland swiddens, nor do the Agta in fact even hunt there. Rather, 90 percent of game is secured in primary forest (Headland, pers. com.). Similar arguments apply to the Batak case and, again I suspect, to numerous other Negrito groups as well.

Further, even if the intrusion of agricultural peoples and the growth of exchange relationships *did* somehow allow the Negrito game resources that figured in such exchanges to thrive, Negrito involvement with exchange may cause them to deplete *other* kinds of resources that are *not* traded but that are nevertheless important to their subsistence. Thus Batak desires to obtain lowland foods and other goods, and hence to have access to the lowland traders who can provide these goods, lead them to locate their "permanent settlements" within reasonable walking distance of lowland communities. Although the Batak still remain strikingly mobile between various residential locations (see Eder 1984), the settlements themselves are frequently inhabited. Not surprisingly, traditionally important riverine foods such as fish and mollusks are badly depleted in the vicinity of Batak settlements, and important game such as wild pig, jungle fowl, and gliding squirrels can scarcely be found there.

To return to the terms of Peterson's (1978) hypothesis, even were it true that a Negrito might spend *less* time collecting and exchanging certain forest products for domestic carbohydrates than if he produced these carbohydrates himself, he might, in consequence, have to spend *more* time producing some item he needed (or simply suffer the lack of that item). The Batak speak nostalgically about the relative abundance of fish and game in the interior, away from their fields and settlements, but their involvement with exchange relationships prevents them from foraging in the more remote parts of their territory as much as they otherwise might. Because food purchases or added foraging efforts during periods of settlement residence do not compensate for the reduced availability of forest and riverine foods, diet is visibly poorer at such times than at interior forest camps. Hence, depletion of nonexchange subsistence resources must be judged a significant economic cost of Batak involvement with lowland "exchange relationships."[1]

Trade relationships with wider Philippine society also have broader social costs to Negrito populations. Anthropologists have long commented on the socially fragmenting consequences of trade for hunting-gathering populations (e.g., Murphy and Steward 1956). Morris (1982) has recently argued that the small co-resident groups of the Hill Pandarum in south India reflect the demands of trade in forest products rather than the demands of a subsistence hunting-gathering economy. Elsewhere (Eder 1983) I develop a similar argument for the Batak. Also widely observed are the interpersonal tensions and conflicts associated with reliance on external trade (e.g., Nietschmann 1973; Savishinsky 1974). Arguments between husbands and wives about the purchase of trade goods, and between neighbors about the sometimes competitive display of these goods, testify to the impact of trade on Batak life. Such interpersonal tensions likely affect other Negrito groups as well.

Finally, I want to call attention to what can be termed the "deculturating" potential of Negrito/lowland Filipino exchange. This arises in part because of the socially fragmenting influence of trade just noted. Much deculturation occurs, however, because as economic life comes increasingly to revolve around exchange and other relationships with lowland Filipino society, elements of traditional culture that are irrelevant to such relationships fall by the wayside. Much *acculturation* occurs as well, as when indigenous musical instruments are abandoned for factory-produced guitars. Negritos, however, seem particularly susceptible to loss or abandonment of indigenous beliefs and practices *without* functional substitution, with a resulting net loss in cultural inventory. If culture is indeed "adaptive," as it is often said to be by anthropologists, then any such deculturation is potentially threatening to individual well-being. Certainly many Negritos are extensively deculturated; many factors are admittedly involved, but I believe that the reorganization of traditional life entailed by the intensification of trade relationships is one of them.

Some Historical and Evolutionary Considerations

The notion of "deculturation" attending the development and intensification of Negrito/Filipino exchange relationships brings us to the second shortcoming of Peterson's (1978) analysis: her failure to locate such exchanges in a satisfactory historical or evolutionary framework. Peterson has observed Agta-Palanan relationships at a particular point in time but writes as if the adaptive advantages to the Agta of their "tandem specialization" with the Palanan were timeless (1978:347–348). As long as the Palanan do not upset the applecart by raising more domestic animals (1978:347), and as long as the Agta do not "increase competition" by expanding their own agricultural efforts (1978:344), the system will be "relatively stable through time" (1978:347)—unless change is imposed by "deforestation by logging companies" or by "government intervention, either developmental or military" (1978:347).

In fact, however, change and adaptive difficulty in Philippine Negrito groups predate logging companies and "the government." For centuries, outside agricultural populations—what has become lowland Filipino society—have expanded against Negrito populations, and in areas where Negritos once thrived, only remnant groups remain. For example, according to popular sources, about 14,000 Negritos inhabited Panay in 1849. But by 1900 there were less than a thousand Negritos, and today only a "handful" of uncertain ethnicity remains (Coutts et al. 1981:109). Similar histories of the demographic and cultural effects of contact have been repeated throughout the Philippines, albeit during different time periods; historical factors and comparative geographical isolation probably explain why a few Negrito groups (e.g., the Agta) have persisted in relatively pristine form down to the present.

I would like to make two points about this expansion of non-Negrito peoples against Negrito groups. First, it likely did not occur (for it does not occur today) in the absence of regular social and economic relationships, including exchange relationships, between members of the populations in question. Indeed, trade has likely been an early, regular feature of the "domestication" of Philippine Negrito groups, helping to make Negrito resources available to outside peoples and to bring the Negritos themselves under wider orbits of political control (see papers by Cadeliña and Rambo, this volume). Once exchange relationships are well established, less complementary relations follow—asymmetrical labor relations (note that Agta do not hire outsiders) and land expropriation (note that Agta do not steal outsiders' land). Peterson has simply observed a case of Negrito/non-Negrito interaction at a relatively early (and benign) stage in this evolutionary process.

Second, at more advanced stages of this admittedly greatly simplified evolutionary process, it becomes less appropriate to speak of two distinct populations locked in economic interdependence but each otherwise distinctive in language, culture, and social boundaries. What emerges instead is a single, internally specialized population in which hunter-gatherers do not simply exchange with farmers but fill an occupational niche in a wider economy. If many of the remaining Philippine Negritos will survive at all, I think it will be on these terms—i.e., as members of specialized production units within a single, wider society.

Fox (1969), reevaluating the status of south Asian hunter-gatherers, provides a provocative model for envisioning their condition that is also useful in the case at hand. If there is any endpoint, other than extinction, in the evolutionary trajectory traveled by Philippine Negrito groups, it is probably closer to Fox's model than to Peterson's. According to Fox (1969:141–143):

> Rather than being independent, primitive fossils, Indian hunter-and-gatherers represent occupationally specialized production units similar to caste groups such as Carpenters, Shepherds, or Leather workers. Their economic regime is geared to trade and exchange with the 79 more complex agricultural and caste communities within whose orbit they live. Hunting and gathering in the Indian context is not an economic response to a total, undifferentiated environment. Rather, it is a highly specialized and selective orientation to the natural situation: where forest goods are collected and valued primarily for external barter or trade, and where necessary subsistence or ceremonial items—such as iron tools, rice, arrow-heads, etc.— are only obtainable in this way. Far from depending wholly on the forest for their own direct subsistence, the Indian hunters-and-gatherers are highly specialized exploiters of a marginal terrain from which they supply the larger society with desirable, but otherwise unobtainable, forest items such as honey, wax, rope and twine, baskets, and monkey and deer meat.
>
> From this perspective, Indian hunting-and-gathering groups are viewed as marginal economic specialists for traditional Indian civilization. As a result of this role, their external interaction

must be seen as economic enclavement within the larger society. This same enclavement mandates important internal social modifications. That is, the social organization of these groups is transformed to meet the expectation of collection and exchange with the outside world. This view-point fits Indian hunter-gatherers into a general social framework identifiable as the type process of Indian civilization: the localization within bounded kinship or "ethnic" groups of occupational and craft specializations.

When hunter-gatherers are incorporated into wider societies in this fashion, what happens to their social identities and to the ethnic boundaries by which they were conventionally differentiated from other peoples? Do social boundaries persist at all, other than those entailed by occupational and economic differences? Assuming a modicum of endogamy is practiced, distinctive biophysical characteristics alone ensure a positive answer to this last question, in the case of Negrito populations. Cultural differences, to be sure, will persist as well, but—and based only on my observations of the Batak case—the *relevant* differences undergo a subtle transformation as incorporation into a wider society proceeds. First, the cultural identity of the hunter-gatherers comes increasingly to be defined in terms of those practices and values related to exchange. Thus a Batak may come to be someone who collects and exchanges a characteristic set of forest resources, whatever else he or she may do. Second, and reflecting this first transformation within the same wider social system, the ethnic boundaries separating *different* hunting-gathering groups (and even those boundaries separating hunting-gathering groups and other kinds of indigenous peoples) become blurred, for numerous different indigenous peoples, after all, collect and exchange forest resources. (Elsewhere in this volume, Cadeliña similarly argues that Batak/Tagbanua ethnic differences have been muted by their common resource position *vis-à-vis* lowland Filipinos.)

Thus, and particularly as increasing intermarriage dilutes distinctive racial features, my own image of the future of the Batak is that they will be identified less as Batak and more as a kind of generic tribal population. As such, they will be demarcated not by any distinctive cultural marker (e.g., use of the bow and arrow) but by the same set of diagnostic social and economic factors—marginality, poverty, illiteracy, limited cultural inventory—that generally characterize evolving Philippine tribal populations and for whom vaguer ethnic designations are more appropriate (e.g., in Palawan, terms such as "tribos" or "Tagbanua").

Swift (1978), surveying the current status of small-scale societies, including hunter-gatherers in Asia and the Arctic, has aptly summarized such transformations by placing them in the wider context of class formation:

The process by which isolated small-scale societies are incorporated as marginal components of a larger universe is usually also the process by which class formation is started. The commercialization of previously subsistence economies leads to the emergence of new and more permanent economic and social inequalities; the new institutions and roles that are created to mediate between the small society and the larger often become the institutions of a new class system. As a result, the problem of the marginal society begins to become a problem of class as much as ethnic or cultural identity, although it may continue to be perceived and formulated solely as the latter. (Swift 1978:13–14)

The Question of Negrito Farming

The approach taken in this chapter also sheds new light on the question of why Philippine Negritos have generally failed to move into full-time farming. Philippine Negritos *do* farm; indeed, they characteristically farm to such an extent that they were specifically excluded from consideration in the *Man the Hunter* volume (Lee and DeVore 1968:17). But with few exceptions, farming is strictly a part-time activity, one of a characteristic set of opportunistic subsistence activities—including trade and hunting-gathering itself—by which "hunter-gatherers" in this part of the world are typically known (e.g., Griffin 1981:32; Endicott 1974:32–36). At first glance, it might appear that Negritos are traveling some sort of evolutionary trajectory from full-time hunting-gathering to full-time agriculture. Indeed, it is often assumed that particular Negrito groups are "in transition" between these two subsistence adaptations, an assumption visible in the title of Warren's (1964) monograph on the Batak or Rai's (1982) dissertation on the Agta.

My opinion, however, is that part-time farming represents a fairly stable adaptation in its own right. If any such evolutionary trajectory is present, it must have a very long time-frame indeed. Philippine Negritos have been in contact with agricultural populations for at least several centuries; they have probably farmed part time for much of this period, living in environments where full-time farming is possible (Hutterer 1983:173–174). Although population pressure, resource depletion, and the incursions of outside populations have already, in some cases, diminished the importance of hunting-gathering to the characteristic Negrito subsistence stance, there are other possible outcomes than full-time agriculture. A tendency to see Philippine Negritos as more "in transition" than other kinds of people is probably more an artifact of anthropological reliance upon pure-type models (pure hunter-gatherers, pure agriculturalists) than upon any actual tendency for mixed adaptations to be more likely to change over time than pure ones.

The question of Negrito farming is not unrelated to the question of Negrito exchange relationships discussed earlier, for an ancillary argument sometimes adduced to support an emphasis on the functional benefits of trade to Negrito populations is that Negritos themselves *prefer* trade to the alternatives, particularly agriculture. The basic reasoning is that Negritos need, in any case, certain goods from the outside world (e.g., agricultural foods, clothing) and that for a variety of reasons having to do with the "fit" between collection and sale of forest products and hunting-gathering of forest resources for subsistence (e.g., because it allows continued mobility), Negritos find trade a more congenial way of getting these goods than they do from such relatively sedentary occupations as farming.

Writing of the Batek, a Malaysian Negrito group similar in many respects to the Batak and the Agta, Endicott (1979) has argued persuasively along these lines. His argument has considerable intuitive appeal and versions of it have been applied to Philippine Negrito groups as well (e.g., Headland 1981). Most (Malaysian) Negritos, Endicott says,

> intensely dislike agricultural work and will do it only when it is impossible to earn a living any other way. They much prefer collecting and selling rattan to agriculture because it permits them to live in the cool forest, to move around whenever they like, to hunt and fish, and to collect their reward in cash, which they vastly prefer to a yield of food alone. . . . This preference is widespread among Negritos and is a serious impediment everywhere to attempts to settle them. (1979:184)

Not only do Negritos dislike agricultural *work*, writes Endicott, but because of "their preference for the coolness of the forest to the heat of the clearings," their "intense dislike of . . . living in one place for an extended period," and their "avoidance rules between kin which severely limit who may live with whom," Negritos find the house layout and domestic arrangements of Malay-style agricultural villages "almost the exact opposite" of what "they look for in a living place" (Endicott 1979:184–187). Later, Endicott (1979:187) also reveals that farming is not a particularly productive or reliable occupation in the area inhabited by the Batek. In the absence of quantitative data on costs and returns for different subsistence activities, one is hence uncertain whether simple economic calculation, or broader cultural preferences divorced from such calculation, in fact underlie Batek subsistence choice. In any case, the cultural preference portion alone of Endicott's argument is persuasive.

But such arguments need closer inspection. Again, I do not deny the generally desultory state of Philippine Negrito agriculture, nor do I deny that Negritos talk (and act) as if they do not like farming work and the farming lifestyle. But I do question whether these facts are adequately accounted for by the claim that agriculture, as a subsistence adaptation, is simply a "poor fit" with traditional Negrito social organization, culture, or personality. For in order to fully appreciate the motivational patterns in question, we must again look at the wider social and economic context in which the failure to farm effectively occurs. If we are to invoke that context to explain why Negritos have become such inveterate collectors of commercially valuable forest products, surely we must consider it as well in accounting for the failure of Negritos to move into full-time farming. My view is that the terms of incorporation of Negrito society into lowland Filipino society do not simply select *for* trade; they select *against* farming.[2]

Let me elaborate. First, I think it is important to recognize that specialization and success in one activity precludes, to a degree, specialization and success in another activity. I argued earlier that Peterson (1978) ignored some significant opportunity costs associated with developing and maintaining exchange relationships. Some of these opportunity costs figure in Negrito farming failure. In particular, and again using the Batak case for illustration, the ongoing concern of an individual Batak to stay in the good graces of his lowland patron/creditor/exchange partner leads him, among other things, to accept offers of employment in his patron's swidden field, performing such tasks as clearing or weeding—the same tasks that badly need doing in his own swidden. (Thus it is not simply that the Batak have failed to move into full-time farming; even their part-time farming is not particularly successful.)

That during the swidden season the Batak may be seen busily laboring in the fields of others while their own fields (or would-be fields) lie unattended is, in fact, commonly cited by lowlanders as indicative of the kinds of personality shortcomings that explain why the Batak have been unable to get ahead economically. In fact, however, many Batak are *obliged* to work in this fashion because they had previously obtained food or other goods from their employers, to be paid for either in forest products or in agricultural labor days. Thus there is an important sense in which a day's work is a day's work and in which the Batak regard a day spent weeding a settler's field to fulfill a debt obligation pretty much the same as they regard a day spent gathering honey for the same purpose. At the very least, we must recognize that "hunter-gatherer/farmer exchange" and "labor relations" are not easily separated aspects of the Negrito relationship with wider society, but may be intimately related. Further, we must appreciate that Negrito involvement in exchange and labor relationships seriously *inhibits* their *ability* to become more successful farmers, whatever their *inclinations* may be. Writing of the Tagbanua, an indigenous and largely *agricultural* people elsewhere in Palawan, Conelly (1985) emphasizes how some of the time spent collecting and selling forest products might alternatively be devoted to clearing larger fields and to better maintaining those fields. More than the opportunity costs of exchange relationships with lowlanders keep Negritos from moving into full-time farming, however. I do not think it is unreasonable to argue that *lowlanders themselves* do not *want* Negritos to move into full-time farming. I do not claim that conscious motivations along these lines necessarily exist, but I do think it is important to recognize that particular lowlanders, and the wider socioeconomic system, benefit economically from the present subsistence orientation of many Negrito groups. Consider that any significant movement by Negritos into full-time farming would (1) increase competition for agricultural land in certain Philippine frontier areas already filled with land-hungry settlers, and (2) decrease the number of marginalized individuals available to collect and sell forest products, thereby raising the costs or decreasing the supply of these products in local, regional, and world markets, and cutting into the profits of those lowland middlemen who presently benefit economically from the trade in question.

Two aspects of the Batak case are relevant here. First, on numerous occasions I have observed lowland settlers, usually those with exchange relationships with the Batak, attempt to intimidate the Batak about their allegedly illegal swidden-making (e.g., with stories about how a certain government forester, having spotted the illegal swiddens from his helicopter, is enroute to make arrests).[3] I have known such efforts to be successful, with the Batak thus intimidated abandoning their swiddens and retreating to the forest for a spell. Such behavior, which may occur during any phase of the swidden cycle, obviously undermines agricultural productivity as much as any aversion to working under the heat of the sun. (The irony of such incidents is that the Ministry of Natural Resources, at least in this part of Palawan, is more concerned with the illegal swiddens of lowlanders than with those of the Batak.)

Second, and taking a historical view of the Batak case, arriving Filipino settlers did not simply displace the Batak from part of their aboriginal territory and then use that land for agriculture; they stole land on which the Batak *had already made agricultural improvements*. The five coastal Batak reservations, set aside by the government with little production during the 1920s for Batak settlement, were all planted (to a degree) in banana and coconut by the time they were overrun by lowlanders. The Batak remember this experience bitterly. In my opinion, the fear that improved land could similarly be stolen in the future, together with continued uncertainty about their land tenure status generally, is a major factor explaining why the Batak have made only tentative efforts to develop their agricultural economy by planting such high value and ecologically sound tree crops as coffee and cashew—crops that would not appear to unduly jeopardize any traditional Batak cultural preference for mobility, independence, or forest living.

These observations, as I have indicated, are limited to the Batak case, but I am confident that similar observations could be made of other Philippine Negritos as well. Here, however, let me simply summarize my argument thus far: There is more to the story of desultory Negrito farming than that "Negritos don't like to farm." The pressures and opportunities of an external social system have a powerful influence on all aspects of Negrito behavior. Just as Negritos *do* collect and sell certain forest resources because an external economy needs them to do so, they *do not* become full-time farmers because that same economy does not need them to—indeed, it does not *want* them to. To anthropomorphize further, wider Philippine society *wants* Negritos to collect and sell forest resources but it *does not* want them to farm. Success in trade and failure in farming are two sides of the same coin—not the coin of the likes and dislikes of Negritos, but the coin of the pressures and requirements of a wider social system. In such circumstances, it should not be surprising that Negritos might attempt to make the best of a bad deal by surrounding their inability to farm effectively with talk of how they do not like farming anyway.

An instructive exception of all of this are the Ayta of western Luzon. The Ayta are a Sambal-speaking population of Negritos, one of a cluster of Negrito populations inhabiting various parts of the Zambales mountains. The Ayta, however, spend little time hunting and gathering; "they are today fully and firmly committed to a subsistence system based on swidden agriculture" (Brosius 1983:139). Indeed, "nearly 100 percent of the total diet is derived from swid-dening" (1983:140), and it is, claims Brosius, the "highly productive" nature of that swiddening that enables the Ayta to "maintain their traditional 'sovereignty' and self-sufficiency" (1983:123). Although I have only visited the Ayta briefly, I am inclined to agree with Brosius's implied assessment that, in comparison with other Philippine Negrito groups, they are pretty well off indeed:

> The degree to which the Negritos within the research area are nutritionally self-sufficient is remarkable. They trade frequently, but always for cloth and clothes, metal, batteries, tobacco, and salt. Though it is a highly desired food, rice makes up only a small percentage of trade volume. Self-sufficiency can be maintained almost indefinitely. It is the source of their strength, independence, and persistence. Their days are not "numbered," as perhaps are the days of other less independent Philippine Negrito groups. Those who would wish to "help" them, as many have, must recognize and come to terms with this self-sufficiency. (1983:144-145)

At first glance, the Ayta case seems counter-intuitive. Here are Negritos who have apparently maintained their physiological well-being and their racial and cultural distinctiveness despite giving up, centuries ago (for reasons Brosius explores), a hunting-gathering lifeway for an agricultural lifeway. Worse, they even appear to enjoy it. Could it be that the Ayta thrive because they embarked on an evolutionary trajectory leading not to ever more debilitating and dependency-breeding exchange relationships but to full-time agriculture? I think we must consider the possibility.[4] I fear that a new romanticism has come to influence analyses of Southeast Asian hunting-gathering populations, the old romanticism about pure hunter-gatherers having given way to a new but equally unrealistic glorification of generalized foraging, multidimensional subsistence strategies, and exchange as what the sensible and well-adjusted modern day hunter-gatherer does. It certainly seems odd to me that anthropologists who work with *agricultural* peoples continually emphasize the importance of greater attention to subsistence production and to the nutritional, socioeconomic, and other dangers of cash cropping (i.e., specialized production for exchange), while anthropologists who work with *hunting-gathering* populations, at least in Southeast Asia, do not voice similar qualms. Yet as the experience of the Batak—"generalized foragers" par excellence—illustrates, there are important parallels. I think we should be more attuned to the possibility that whatever Negritos may be doing at the moment may not necessarily be in their own best interests, however difficult it may be to define those interests.

Acknowledgments

I studied the Batak for two months during 1975, supported by an Arizona State University Faculty grant-in-aid, and between August 1980 and December 1981, supported by a sabbatical leave from Arizona State University and a grant-in-aid from the Wenner-Gren Foundation.

I would like to thank Tom Headland for his thoughtful comments on an earlier draft of this chapter.

Notes

1. Quantitative data showing that foraging returns are lower and diets less diverse at Batak settlements than at Batak forest camps are discussed in Eder (1987).
2. Note that these different perspectives imply different agenda for "development" work with distressed Negrito populations. If Negritos eschew full-time agriculture because it fits poorly with their traditional

cultural arrangements, we are understandably loath to intervene with agriculturally oriented programs. If, however, contemporary Negrito subsistence strategies are guided more by the demands of an external socioeconomic system than by any internal wisdom of their own culture, more aggressive development efforts seem to be justified.

3. "Illegal swiddens" are those made by squatters on public lands not yet officially released by the government for agriculture.

4. This is not to say that all Philippine Negritos should necessarily be encouraged to move into full-time agriculture; local opportunities, constraints, preferences, and traditions must always be carefully examined. Further, it is true that some Philippine Negritos (e.g., those in Negros) are largely agricultural but are poorly off. But then so are many lowland Filipino farmers, and anthropologists characteristically explain *their* failure to farm more effectively not by reference to alleged Filipino value orientations but in terms of situational economic factors such as land tenure insecurity, shortage of investment capital, or exploitative creditor-debtor relationships.

References

Brosius, J. Peter
> 1983 The Zambales Negritos: Swidden agriculture and environmental change. *Philippine Quarterly of Culture and Society* 11:123-148.

Conelly, W. Thomas
> 1985 Copal and rattan collecting in the Philippines. *Economic Botany* 39:39-48.

Coutts, P.J.F., J.P. Wesson, S. Galvego, and D. Dinopol
> 1981 A summary report of the Fifth Australian Archaeological Expedition to the Philippines, 1980. *Philippine Quarterly of Culture and Society* 9:77-110.

Eder, James F.
> 1977a Modernization, deculturation, and social structural stress: The decline of the Umbay ceremony among the Batak of the Philippines. *Mankind* 11:144-149.
> 1977b Portrait of a dying society: Contemporary demographic conditions among the Batak of Palawan. *Philippine Quarterly of Culture and Society* 5:12-20.
> 1978 The caloric returns to food collecting: Disruption and change among the Batak of the Philippine tropical forest. *Human Ecology* 6:55-69.

1983 Seasonal foraging camps as a basis for inferences about an aboriginal hunting-gathering economy. Paper read at the American Anthropological Association Annual Meeting, Chicago, November 16-20.

1984 The impact of subsistence change on mobility and settlement pattern in a tropical forest foraging economy: Some implications for archaeology. *American Anthropologist* 86:837-853.

1987 *On the Road to Tribal Extinction: Depopulation, Deculturation, and Maladaptation Among the Batak of the Philippines.* Berkeley, CA: University of California Press. Forthcoming.

Endicott, Kirk M.
1974 Batek Negrito economy and social organization. Ph.D. dissertation, Harvard University.

1979 The impact of economic modernization on the *Orang Asli* (aborigines) of northern Peninsular Malaysia. In *Issues on Malaysian Development*, edited by J. Jackson and M. Rudner. Singapore: Heinemann Educational Books. Pp. 167-204.

Fox, Richard G.
1969 Professional primitives: Hunters and gatherers of nuclear South Asia. *Man in India* 48:139-160.

Griffin, P. Bion
1981 Northern Luzon Agta subsistence and settlement. *Filipinas* 2:26-42.

Hart, John
1978 From subsistence to market: A case study of the Mbuti net hunters. *Human Ecology* 6:325-353.

Headland, Thomas N.
1981 Imposed values and aid rejection among Philippine Negritos. Paper presented at the 2nd International Philippine Studies Conference, Honolulu, June 27-30.

Hutterer, Karl L.
1983 The natural and cultural history of Southeast Asian agriculture: Ecological and evolutionary considerations. *Anthropos* 78:169-212.

Lee, Richard B., and I. DeVore (editors)
1968 *Man the Hunter.* Chicago: Aldine.

Morris, Brian
    1982 The family, group structuring, and trade among South Indian
        hunter-gatherers. In *Politics and History in Band Societies*, edited
        by Eleanor Leacock and Richard Lee. New York: Cambridge
        University Press. Pp. 171–188,

Murphy, R.F., and J.H. Steward
    1956 Tappers and trappers: Parallel processes in acculturation.
        Economic Development and Cultural Change 4:335–353.

Nietschmann, Bernard
    1973 *Between Land and Water*. New York: Seminar Press.

Peterson, Jean T.
    1976 Folk traditions and inter-ethnic relations. In *Directions in Pacific
        Oral Literature*, edited by A. Kaeppler and H. Arlo Nimmo.
        Honolulu: Bishop Museum Press. Pp. 319–330.
    1978 Hunter-gatherer/farmer exchange. *American Anthropologist*
        80:335–351.
    1981 Game, farming, and interethnic relations in Northeastern Luzon,
        Philippines. *Human Ecology* 9:1–22.

Rai, Navin K.
    1982 From forest to field: A study of Philippine Negrito foragers in transi-
        tion. Ph.D. dissertation, University of Hawaii, Honolulu.

Savishinsky, Joel
    1974 *The Trail of the Hare: Life and Stress in an Arctic Community*. New
        York: Gordon and Breach.

Swift, Jeremy
    1978 Marginal societies at the modern frontier in Asia and the Arctic.
        Development and Change 9:1–22.

Turnbull, Colin
    1965 *Wayward Servants: The Two Worlds of the African Pygmies*.
        Garden City: Natural History Press.

Warren, Charles P.
    1964 *The Batak of Palawan: A Culture in Transition*. Research Series
        No. 3, Philippine Studies Program, University of Chicago.

CHAPTER 4

A COMPARISON OF BATAK AND ATA SUBSISTENCE STYLES IN TWO DIFFERENT SOCIAL AND PHYSICAL ENVIRONMENTS

Rowe V. Cadeliña

In the Philippines, lowland farmers are migrating to the uplands in increasing numbers, placing themselves in a situation of potential competition with upland tribal populations for natural resources of the area. How do the tribal populations whose subsistence strategies are affected by this situation adjust? This question bears directly on the tribal populations' subsistence styles, especially their utilization and management of increasingly scarce resources. In addition, how does this lead to ethnic differentiation, or the dissolution of ethnic markers? This concerns the process of maintaining or eliminating ethnic boundaries that may help groups preserve control over, or increase access to, various resources. These are questions that interest Philippine anthropologists today.

In this chapter, I will compare two Philippine Negrito groups, the Ata and the Batak,[1] who live in different social and physical environments. Both groups were originally nomadic forest foragers, but they have evolved very different subsistence styles in recent years. Taking into account their existing environments, the differences in their subsistence styles in adjusting to migration of lowlanders to the uplands will be explained, and the theoretical implications of the findings for ethnic diversity examined.

Theoretical Framework

The subsistence style of a particular population largely depends on two factors: (1) the population and the social environment, and (2) the resource condition and the physical environment.

In this chapter, the definition of ethnic groups takes into account the processual and ecological components of ethnic differentiation. Interethnic differentiation is assumed to be a process characterized by systemic interaction between the components of the human and natural ecosystems. The strategies and activities by which various ethnic groups utilize resources in a particular area provide the data for analyzing ethnic relations as a process. This ecological definition of ethnic unit classification is patterned after Abruzzi's (1982) model of "resource partitioning" in a multiethnic community.

The model assumes that ethnic groups establish patterns of social organization and relations evolving a process of resource partitioning for advantageous use of energy and resources (Harris 1964, 1977; Blau 1967; Levins 1968; Abruzzi 1981, 1982). The total population of a region may serve as the analytical or investigative unit, rather than each discrete ethnic group. Unlike sociocultural criteria, which establish an *a priori* definition of ethnic groups, the ecological definition may be based on local empirical criteria of categorization (see Barth 1969:13-15) as indicated by the processes through which ethnic groups interact and utilize the resources in the region. Since the definition is process-oriented with a systemic analytical approach, a framework is provided by which the evolution and dissolution of ethnic boundaries may be analyzed ecologically without disregarding other historical, economic, political, sociocultural, and demographic explanatory factors.

The resource condition refers to the structure, content, and stability of resources in an environment. Basic food resources may be evenly distributed throughout the community or skewed (Jochim 1981). In conditions of patchy or clumped distribution, households or population groups physically closer to resources have a definite edge in utilizing them (Wiens 1976). Such a situation could discourage other groups from tapping such resources; however, through social links, a greater number of people may benefit from them (Endicott 1977).

The content of the resource base, in terms of diversity, can also affect subsistence style. Since the use of diverse resources may be found to be more advantageous than the use of highly specialized ones, in which user options are limited, corrective measures to increase the level of diversity may be implemented through the social system. Trading and other forms of energy transfer may be resorted to (Dirks 1980; Laughlin 1974). In situations of high resource diversity, self-contained communities may be able to develop.

Resource stability may be considered in terms of short-term and long-term fluctuation. Short-term fluctuation is generally in the form of expected or anticipated cyclical decline of products caused by seasonal availability. The resource users have a general idea of the schedule, nature, and intensity of the fluctuation (Colson 1979). Longer-term fluctuation, on the other hand, is characterized by the progressive, irreversible decline of resources, grossly indicated by the negative difference between present and past average production. In both cases, the problem of resource fluctuation can be corrected either through technological or social means. Although the problem of resource fluctuation can be minimized through both technological and social means, the effects of seasonal fluctuation are generally minimized through the social system (Steward 1955), and technological innovation is probably imperative to deal with long-term fluctuation.

To establish an optimal level of stability in the resource supply, an ethnic unit may maximize "the number of links in the food web of the community" (Leigh 1965:777). Such a strategy may be established by fluid rather than rigid interethnic differentiation to facilitate a maximum number of interethnic links. Through these links, niche diversification may be established.

Although differential efficiency in the use of resources may eventually exclude one group from use of those resources—a case of exclusive fitness—the survival interests of all units may be as important—a case of inclusive fitness. Instead of competition, symbiotic relations (mutualism) may be established between groups, as in specialization of ethnic niches where one group provides a product to another which does not produce that product. According to Peterson (1978), this is exemplified by the Agta and the lowland Filipinos in northeastern Luzon. However, this chapter will demonstrate that, on the basis of resource use, description of human relationships in exclusively biological terms is inadequate. Many different factors, including the political and the sociocultural, affect the types of energy use relationships among ethnic groups. Use of biological models, while useful for analogy, are inappropriate for complete description for social systems as they disguise or ignore political, economic, and cultural aspects of the system (Jochim 1981).

Setting of Two Ethnic Groups

The Ata of Negros

The Ata are a Negrito group occupying the southwestern mountain ranges of the municipality of Mabinay, Negros Oriental (Map 4.1), which is about 87 kilometers northwest of Dumaguete City. The area, formerly thickly covered with primary forest, is now completely deforested. A new vegetative succession, consisting of herbaceous vegetation and shrubs, has taken over. In some places, cogon (*Imperata cylindrica*) has never been controlled by farm cultivation. Because of continuous farm cultivation in most areas, however, cogon grasses do not dominate but constitute the major weeds at the farm edges.

As a result of deforestation, the traditional food base of the Ata—consisting of wild game, freshwater food products, wild fruits and vegetables, and wild carbohydrate sources—is now completely gone. Hunting, collecting, and gathering activities have been completely replaced by more sedentary farming. *Kaingin* (swidden) agriculture has given way to plow agriculture by using carabao and cattle. Therefore, the Ata are now completely sedentary.

The Ata, a native upland population, live in a community together with lowland Cebuano-speaking migrants, and now speak Cebuano. These two population groups can be categorized by resource use. Furthermore, two groups of lowland Cebuano migrants are evident in the community: One group is largely involved in producing cash crops such as sugarcane, and the other is engaged in producing subsistence crops and some cash cropping. These two subgroups approximate different life-styles: the former resembles the urban, while the latter parallels the Ata life-style.

Among the Ata, two subgroups are also evident in terms of subsistence techniques: One group utilizes only human energy and the other combines human and animal energy. It will become apparent in a later discussion that these groups differ in the types of products they produce, the production level, and their access to products and productive resources.

The Batak of Palawan

The Batak are still relatively nomadic, regularly shifting residence from swidden sites to river banks. The forest base of the Batak is relatively intact. Their traditional hunting, collecting, and gathering activities still exist, making up more than 50 percent of their total production.

The Batak under study occupy the Kalakuasan area of Barangay Tanabag, which is more than 70 kilometers northeast of Puerto Princesa City, the capital of Palawan (Map 4.2). There are six major Batak villages in Palawan, one of which is the Kalakuasan Batak Village.

Although the Kalakuasan area has been logged, the more rugged part of the terrain was left untouched. Its rough topography protected the area from extensive logging activities. Therefore, this particular area provides the Batak with substantial forest commercial products such as *almaciga* resin (taken from the *Agathis philippininsis* tree), wild honey, and rattan. The Batak live in a more ethnically heterogeneous community than do the Ata. The two general categories of indigenous uplanders and lowland migrants are also applicable to the Batak, but the composition of each category is far more complex.

Culturally and linguistically, the upland population is composed of the Batak and the Tagbanua. Although these two groups demonstrate some cultural overlap, a number of distinctive sociocultural traits distinguish one from the other. The Ata, in contrast to the Batak, constitute the only indigenous upland population group in their community.

The lowland migrant population in the Batak community is highly heterogeneous. Although the Ata community's migrant population is composed entirely of Cebuanos, a number of linguistic groups coming from various parts of the country are represented in the Batak community: the Cuyunon, the Cebuanos, the Ilocanos, the Batangueño, and the Ilonggo, to mention just a few.

One interesting linguistic development is the emergence of one common language among these highly heterogeneous ethnic communities. Tagalog is fast becoming the *lingua franca* for all ethnic groups. As a result of contact and consequent acculturation, multilingualism is also apparent. For instance, the Batak are basically quadrilingual. They speak Tagalog (especially the males), Cuyunin, Tagbanua, and Binatak. The Ata, on the other hand, have completely lost their traditional language and are now monolingual.

Comparison of Ata and Batak Subsistence Styles

For the discussion on subsistence styles, three issues will be dealt with: carbohydrate procurement, protein capture, and material provision. The first and second concern biophysical needs, and the third their social demands.

Carbohydrate Procurement

The Batak and the Ata differ in their carbohydrate food sources. Table 4.1 shows the pattern of carbohydrate consumption of sample household cases in one agricultural year and suggests the way in which carbohydrate foods are procured. The Batak are not as dependent on their farms as are the Ata. While the Batak obtain as much as 60 percent of their total carbohydrate foods (cereal and root crops combined) from their farms, the figure for the Ata is as high as 75 percent. This figure suggests more intense farm work among the Ata, reflecting the complete disappearance of the forest in the Ata community coupled with the upsurge of lowland migrants.

For instance, during the 1983–84 study, the Ata no longer collected wild root crops. In contrast, 10 percent of the Batak carbohydrate diet still included wild root crops, due to the preservation of the forest and traditional foraging activities.

The consumption patterns of bought or exchanged food among the Ata and Batak require closer examination. During the months of scarcity, more cash circulates in the Batak than in the Ata community because the Batak can sell commercial forest products. *Almaciga* resin, rattan, and wild honey constitute the major sources of cash. These cash opportunities are no longer available to the Ata.

Two distinct ethnic groups can be distinguished in Palawan by their roles in
trading transactions. The Batak, who have the knowledge and skills in forest
collecting, serve as procurers. They control that niche with other natives such as
the Tagbanua. The lowland migrant groups take the role of middlemen with the
larger Philippines and world economy by acting as buyers of the forest products.
They are able to control and monopolize the buyers' role because they have
access to enough cash with which to start. The lowland migrants have
attempted to learn the úplanders' collecting skills but without much success and,
until now, at least in the village studied, the Batak still control that activity of
the system.

The relationship between the procurer and the buyer is cash-oriented and
kept on an impersonal level. However, buyers and sellers were observed to grant
a special favor occasionally to their partners as a way of maintaining patronage.

The Batak buy rice from lowland middlemen during the months of scarcity.
No generalized exchange of cereals between the Batak and the lowland buyers of
forest products ever took place during my study. The Batak always paid for the
rice with cash from wage labor or by labor. Closer examinination on the sources
of cereals that were bought and traded or exchanged revealed that about 20
percent of these foodstuffs came from generalized exchange within the Batak
population and with a few Tagbanua. The interhousehold movement of
resources is affected by the developmental cycle stages of the household, the
agricultural cycle, and by the consanguineous or affinal links between
households. The few Tagbanua household participants in the Batak interhousehold food-sharing network were found to be affines of the Batak.

The Ata, on the other hand, do not get as much of their cereals from the
lowland Cebuano migrants. This does not mean, however, that there are fewer
intergroup transactions than among the Batak, as the gross percentage
difference is misleading. Therefore, a number of points need clarification.

- The Ata, unlike the Batak, have gone a long way from their traditional
  carbohydrate procurement system. The present Ata agricultural system
  is similar to that of their lowland Cebuano counterparts, although the
  Cebuano farmers may produce a little more than the Ata because of
  acreage differences and other inputs for which the Ata may be unable to
  invest. Still, the Ata manage to get most of their cereals from their own
  farms.
- The similarity of the planting schedules of lowland Cebuano and Ata
  farmers means that their lean and abundant months coincide, making it
  difficult for the Ata to turn to Cebuano farmers at times of cereal scarcity. Also, the Ata have adopted second cropping, the products of which
  are harvested right before the months of scarcity. This provides a buffer
  against hunger. There is no second cropping among the Batak.

- The Ata plant more root crops than the Batak, known as "famine" or *lagutmon* (hunger) crops to the Ata, which are harvestable during famine months. These crops allow the Ata to draw their carbohydrate supply from their own farms; hence, the lower need for obtaining this food commodity from the lowland Cebuanos. The percentage difference between the Batak and the Ata on this aspect is slight, only 5 percent. Nevertheless, the difference between the way the Batak and the Ata produce their carbohydrate supply could have systematically yielded this difference.

- The socioeconomic transactions between the Ata and the lowland Cebuanos are more intense than those of the Batak and their lowland partners. For the Ata, the transaction tends to be more personalized and cash does not serve as the only means to move carbohydrate foods from one group to another. While only 20 percent of the bought, traded, or exchanged carbohydrate foods move through generalized exchange among the Batak, it is three times as much (60 percent) among the Ata. The lowland Cebuanos originated 90 percent of the carbohydrates in generalized exchange, and the rest came from other Ata households. The reverse was true among the Batak—more than 90 percent came from other Batak households.

- More generalized sharing seems to occur between the Ata and the lowland Cebuanos than between the Batak and their lowland trading partners. A closer examination showed an increasing incidence of intermarriage between the Ata and the lowland Cebuanos. Affinal ties provide the intimate link between these ethnic groups, allowing generalized exchange to take place. In 1968, about 40 percent of the Ata couples were intermarried; the figure is now closer to 60 percent (Oracion 1983a). Among the Batak in the Kalakuasan Village, not a single Batak is married to a lowland migrant.

- In general, there is a lower incidence of interhousehold generalized sharing of carbohydrates among the Ata households than among the Batak. Among the Ata, the increasing number of lowland Cebuano affinal relatives provides them with an integrated multiethnic social group in which generalized sharing can take place. Considering that all Ata households are virtually homogeneous in their economic conditions and capability, interhousehold generalized sharing can hardly take place on an inter-Ata household level since they are more or less equally faced with difficulty during the period of scarcity. Although the developmental cycle stage differences between households may allow some of the Ata households to participate in interhousehold generalized sharing, their lowland Cebuano extended affinal kin provide more convenient and effective generalized exchange partners. Oracion's (1983b) study shows that production by lowland Cebuano migrants is superior to that of their Ata counterparts. Consequently, generalized sharing is more likely to take place between an Ata and a lowland Cebuano affinal relative than between an Ata and another Ata affinal relative.

The data suggest that the changes taking place in the personnel involved in the generalized exchange of carbohydrate food among the Ata constitute an adaptation to the increasing scarcity of carbohydrate resources among Ata households and the increasing number of lowlanders coming into the Ata community. The lowland Cebuanos' advantageous position (i.e., amount of land owned, terrain, and perhaps technical knowledge of intensive agriculture) allows them to produce more. The Ata are able to benefit from Cebuano productivity if effective social links are established. Marriage is one such link.

Why do lowland Cebuanos allow intermarriage? Recent data show that the Cebuano mates of the Ata are offspring of parents who migrated in the late 1960s, coinciding with the period when land was becoming difficult to obtain. The Ata who are now settled have their own land, which attracts landless lowland Cebuano migrants. Consequently, intermarriage has probably become a means of acquiring land. In the past, intermarriage seems to have been a matter of personal convenience between Cebuanos males and Ata females.

Protein Capture

Activities are deliberately planned to get protein-rich side dishes to go with carbohydrate foods (*pangsuda* among the Batak and *sud-an* among the Ata). At present, the Batak generally associate *pangsuda* with wild meat and fish (freshwater and marine) while the Ata usually associate *sud-an* with salted or dried fish. Also, domesticated livestock such as poultry and swine have prominent roles in the protein provision among the Ata.

Among the Batak, protein is largely derived from wild animals including wild pig, gliding squirrel, wild peacock, scaly ant-eater, porcupine, wild cat, lizard, Palawan bear cat, monkey, selected birds, and snakes. Both freshwater and marine organisms such as fish, crustaceans, and mollusks are tapped. The Batak, unlike the Ata, have access to a marine fishing ground only 5 to 7 kilometers from their village. The Ata, on the other hand, are about 60 kilometers from the nearest coast; consequently, they are completely ignorant of the resources available in the marine ecosystem. The Ata have completely lost their hunting resource base, and the disappearance of fish from their major streams due to poisoning has also completely eliminated traditional fishing. Hunting and fishing are only a memory; in fact, the present generation mostly reported that they only knew of this activity from their elders' tales.

Table 4.2 shows the sources of protein for the Batak and the Ata. The Batak tend to be more self-sufficient in protein supply and to have more control over it than the Ata. The Batak still control this niche, especially now that they have adopted the locally made muzzle-loaded shotgun. It has tremendously improved their success rate but has put the wild pig in danger of being overhunted. The lowland buyers of forest products and rice-sellers to the Batak derive a good part of their protein supply from the Batak. Ironically, while the Batak control the niche that provides their protein supply, the lowlanders still control the price of wild meat, a situation similar to that in the sale of commercial forest products. Consequently, the meat buyers (middlemen) control the cash supply in the community. In the context of protein use, the meat buyers (lowland migrants) are not distinguished according to their ethnolinguistic orientation. They are simply grouped by the Batak as meat consumers or buyers.

The Ata, on the other hand, are largely dependent on the lowland Cebuano traders for their protein but, like the Batak, do not control the price. While the roles of the Batak and the Ata in protein provision are reversed, there is no corresponding reversal in the roles of price controllers.

The lowland migrants control and dictate the price in any cash transaction, whether the indigenous population involved are protein buyers or sellers. Natives such as the Ata and the Batak do not deal with a specific lowland ethnolinguistic group but with a solid front, the combined force of different ethnolinguistic groups from the lowland communities in the Philippines. These groups constitute one whole unit of *cash* controllers with whom the *native* producers must contend. In the context of resource partitioning, the native tribal populations have increasingly lost control over their own domain by being systematically deprived of benefits from the system. Hence, as the native uplanders move toward a cash economy, their control over the provision of carbohydrates and protein diminishes.

Gradually, however, the native uplanders might be able to regain minimal control by becoming self-sufficient in both protein and carbo-hydrate provisions. For instance, the Ata increasingly keep their own domesticated livestock. They could control their protein supply if the livestock were to be kept solely for domestic consumption, which would also give the individual consumer more freedom. The problem of accelerating loss of control remains when livestock are raised for sale only. If the native uplanders could be encouraged to raise live-stock primarily for their own supply and secondarily for cash, control might be restored.

In addition, the Ata, through their increasing intermarriage with the lowland Cebuanos, have provided themselves with a channel through which protein food can be obtained by means of generalized exchange. Generalized exchange provided 30 percent of all protein not produced by the Ata themselves.

Material Provision

Tools, clothing, housing facilities, and kitchen utensils constitute the major material items (Merecido 1983). While the Batak have retained some of their traditional tools, the Ata have completely replaced theirs with those purchased or bartered from the lowland Cebuanos.

The increasing need by the Batak to provide more meat for sale despite the declining stock of wild game demands tools that improve their catch efficiency. As the wild game supply dwindles, the distribution also becomes highly patchy or clumpy, reducing the likelihood of chance encounters when using only their traditional technology. For instance, the use of bow and arrow in the ambush type of hunting was once quite successful. Today, the chance of success in ambush hunting is almost nil. At present, homemade muzzle-loaded shotguns are shared among the Batak, the Tagbanua, and the lowland migrants. Sometimes arrangements are made by which the lowland migrants provide the weapon and ammunition to the Batak in exchange for a share of the catch. In some cases, the Batak serve as guides to hunting sites for other groups in return for a share of the catch, and food during the hunting trip. In other cases, Tagbanua or lowland migrants may be invited to join in the annual collective hunting activity known as *sagbay* (R. Cadeliña 1982).

In the Ata community, farming tools (especially plows and draft animals) are the items most often shared. Lowland migrant Cebuanos who are not engaged in cash cropping provide the Ata with an opportunity to use the tools and animals for the Ata's growing needs in sedentary farming. Since the Ata still cannot afford to own these farming implements and work animals (due to cash-capital constraint), they have had to make arrangements for sharing, as follows (Oracion 1983a):

- When an Ata uses a lowland Cebuano's carabao and plow to cultivate his own field for a number of days, he also works for the same number of days on the latter's farm. This usually takes place when the Cebuano farmer has resources but lacks the needed manpower.
- The Ata gives a 30 percent share (although the rate varies) of his farm products to the lowland Cebuano from whom he borrowed the carabao and plow.
- When an Ata uses a lowland Cebuano's carabao and a plow, he cares for the carabao in return. The owner has the right to use the carabao and plow for his own field any time he needs them.

Among the Batak and the Ata, there is an increasing desire for material items such as clothing, housing, and kitchen utensils used by the lowland migrants. Both the Batak and the Ata can obtain these resources through barter, purchase, and as gifts from lowland friends, tourists, and researchers. Since tourism is prevalent in Palawan, the Batak enjoy more "gift" opportunities than the Ata, in whose area there is no tourism.

A study by Merecido (1983) revealed a high intensity of material flow, including medical resources, from the lowland migrant Cebuanos to the Ata (F. Cadeliña 1983). To their mutual benefit, the subsistence farming Cebuano exchange materials and information concerning folk medicine (both herbal and nonherbal) with the Ata. The process of exchange is either purposive or accidental and, unlike other resources, is generalized. F. Cadeliña (1983) notes that accidents or other emergencies generally lead to the unsolicited sharing of materials and information, usually motivated by altruism. Among friends or affinal relatives, materials or information on medicine may be voluntarily sought. In addition, modern medicines are often donated by government or private agencies. The Batak also receive medicinal donations.

Ethnic Boundaries

Although the ethnic groups in the Ata and the Batak communities have maintained some of their sociocultural markers, they have nevertheless succeeded in establishing both fluid and rigid relationships with other ethnic groups living outside each community.

The Batak

As noted earlier, the social composition of the Batak community is heterogeneous, including Tagbanua and lowland migrants of many ethnic origins. Batak and Tagbanua niches overlap, but this does not lead to resource competition, mostly because the Tagbanua tap lowland resources, while the Batak prefer to use upland forest resources and turn to lowland resources only as supplements. Conversely, although the Tagbanua still have the same access to forest resources as the Batak, Tagbanua forest-utilizing activities are much less intensive than those of the Batak. This represents a shift of economic interest by the Tagbanua, resulting in wider freedom for the Batak in the use of forest resources, with less demographic pressure.

In temporal terms, the intensity of the Tagbanua's use of upland forest resources fluctuates. There are months when the use of forest resources are left solely to the Batak and others when the Tagbanua and Batak share these resources. For instance, wild pigs are sought by the Tagbanua during the months when wild fruits are ripening (August to September) and the best meat can be obtained. Forest commercial products are also tapped a little more intensively by the Tagbanua during the lean months. Although the need timing of both groups for forest resources coincides, competition between the two groups is kept at a minimum because there is only partial overlap in the use of resources.

In a purely competitive situation, each group tries to eliminate the other from the niche (see Hardin 1960). Among the Batak and the Tagbanua, a number of cooperative ventures in collecting and hunting take place. Tools and resources are shared. The Batak borrow the highly valued muzzle-loaded shot-guns of the Tagbanua. In turn, the Tagbanua *almaciga* collectors depend for their resin on trees máintained throughout the year by the Batak. The loose control of upland resources by the Batak and Tagbanua allows them to meet their respective needs through implementation of multiple shifts in resource use from one base to another for optimal benefits. Thus, the boundary between the Tagbanua and the Batak in the use of forest resources can be described as highly fluid.

Overlaps in resource use between the Batak and the Tagbanua, along with their collaborative activities, reduced competition, dissolved Batak-Tagbanua differentiation, and led to more frequent interaction between them. Most of the Batak speak Tagbanua, and both speak a common language—Tagalog. Some intermarriage has occurred. However, the Tagbanua are categorized by the Batak as an intermediate group living somewhere between the areas occupied by the Batak and the lowland migrants. For instance, some Tagbanua now keep draft animals such as cattle and carabao. Such a perception coincides with the way resources are partitioned between these groups.

At present, not much lowland resources is exchanged between the Batak and the Tagbanua. When there is exchange, it is only incidental and apparently not pursued in a systematic manner by the Batak. The Tagbanua's access to, and control over, lowland resources is not perceived by the Batak as a lucrative source from which to earn part of their living. The Batak would rather draw resources directly from the lowland Filipino migrants.

The Batak treat the lowland migrants as one: "people from the lowland." The lowland migrants themselves have dissolved most of their interethnic sociocultural differentiation. Although they still speak their original languages, they tend toward a pan-linguistic type.

The boundary between the Batak and the lowland migrant population can be described as rigid. Due to the disparity in the skills of the Batak and the lowland migrants, niche overlap has been at a minimal level. That which occurs is usually through mechanisms serving as "gates" through which resources are exchanged.

Trade relations between the Batak and lowlanders have already been discussed. Just as the Batak are dependent on the lowlanders for dealing with the outside market, lowland migrants are unable to successfully obtain commercial forest products without help from the Batak. Although lowlanders have directly tried to engage in collecting, the skills in the exploitation of upland forest resources that the Batak take years to learn are not easily transmitted to lowland migrants.

Only a few Batak have succeeded in establishing ceremonial ties with the lowland migrants such that special favors and privileges are granted to each other. Such privileges take the form of special favors (in kind) granted to the Batak over and above the usual payment for forest products in cash transactions. However, even when freer access to noncash resources are provided by a lowland migrant to a Batak, the buyer's role is never shared.

The rigidity of the resource use boundary between the Batak and the lowland migrants has been perpetuated by the absence of intermarriage between these two groups. The Batak see themselves as participants in the collector-seller dyad, but as separate from the lowland migrants.

The Ata

The Ata constitute the only indigenous upland population living in the community and interact only with the Cebuano. The physical appearance of the Ata and the lowland Cebuano clearly distinguishes them as two separate ethnic groups. Linguistically, however, the Ata and the lowland Cebuano now belong to a single group. The Ata of today have completely lost the language they claim to have spoken before and now speak only Cebuano, with an Ata accent.

Resource use boundaries between the Ata and the Cebuano vary, depending on the resource base of the Cebuano. The resource use boundary between the Ata and Cebuano subsistence farmers is highly fluid and sometimes no longer discernible. While there are fixed physical boundaries established between Ata and lowland Cebuano farms, resources move freely between them in intense generalized exchange. Cash may sometimes be involved, but transactions are largely informal. It is not unusual to find members of both groups pooling their animal and personnel resources, working collectively by turn from one farm to another.

A different picture can be seen when the Ata interact with the lowland Cebuano engaged in commercial cash crop agriculture. Resource exchanges between them, which are highly formal, generally involve cash.

Their superior rates of production and access to economic resources allow the Cebuano to consolidate their already firm control of resources and to gain ascendancy over the Ata. One factor is that during the lean months the food supply of the Ata usually runs out faster than that of the Cebuano. Thus, the Cebuano are a source of food for the Ata, tapped through wage labor and borrowing at a specific rate of interest. This allows the Cebuano to advance loans during the lean months, to their own advantage. The other factor is that the limited land area worked by Ata households makes a great part of their labor force idle for much of the year. This excess labor force is tapped by Cebuanos during the months when labor is needed on the latter's farms. While the Ata maintain a labor supply for seasonal needs, the Cebuano are the major source of demand for such supply and dictate the rate at which labor is compensated.

When the lowland Cebuanos first came to the Ata community, pure competition for resources seems to have taken place. Under these conditions, it might be predicted that the less effective competitor (in this case, the Ata) would be completely eliminated from the area (Abruzzi 1982; Gause 1934; Hardin 1960). This has not occurred because, through extension activities, private agencies provided protection to the Ata so that the deleterious effects of a true competitive system were minimized. The intervention of private agencies has helped the Ata to maintain a hold on the last of their land despite stiff competition from the Cebuanos. The Ata have established rights over their land and have transformed their role in the community from that of mere "takers" of resources from the lowland Cebuano to that of "providers" as well. Hence, the complete elimination of the Ata from the system has not taken place.

Today, the Ata and the lowland Cebuano (particularly the subsistence farmers) vary only in terms of their relative productive efficiency rather than in absolute terms whereby the Ata are completely subservient to the Cebuano. An equalization process in terms of access to opportunities seems to have taken place between them. Abruzzi (1982) predicted that in such situations a system would evolve whereby a portion of the economic system would be restricted to one group and the other portion to another in a complementary manner. This has not occurred between the Ata and the susbistence farming Cebuano but does seem to exist between the Ata and those lowland Cebuano engaged in the production of commercial cash crops. A distinct demarcation of resource use boundaries exists between them, such as the following:

- The schedule of Ata participation in the niche of the cash crop producing lowland Cebuano farmers is well defined.
- The role of the Ata participation is also well defined.
- The life-styles of the Ata and the cash cropping Cebuano are completely different, in contrast to the parallel life-styles of the Ata and the subsistence farming Cebuano.
- The exchange of resources between the Ata and the cash cropping Cebuano tend to be highly formalized and cash-oriented.

The apparently rigid boundary existing between these resource users is preserved by the nature of the niche each one is dealing with, rather than by strong sociocultural sanctions. Although they do not participate in the total technical operation of the industry, the Ata provide an important requirement in the form of labor, so they are not entirely excluded from the cash cropping niche. On the other hand, the Ata land is not really profitable for the sugar industry. Therefore, the Cebuano sugar farmers have no incentive to go beyond their own boundary and the resource use boundary is still apparent.

Definition of Resource Use

Resource use can take place on two different levels: the primary and the secondary. The Batak and the Ata may use (1) land or other natural resources from which products may be derived, which are considered primary resources; and (2) secondary products generated from the primary resources.

The manner differs by which primary and secondary resources are controlled and tapped by each ethnic group. Apparently the degree of differentiation varies as the nature of resources tapped changes. In the case of basic productive resources or primary resources, where the ownership of land is legally defined, the boundary is highly rigid; it belongs to members of one group or another.

In the Ata community where legal ownership of land is today well defined, the land-use system of the Cebuano and Ata is fixed. Land resource use boundaries are rather rigid. In the Batak community, the forest resources are available to all groups, contingent upon their ability to use them, and for as long as the use of the resource does not directly jeopardize each other's usufructuary rights to areas that were once occupied by the Batak. However, a number of cases were observed during my 1980–81 fieldwork when the usufructuary prerogative of a Batak was given up to people from another ethnic group in the area. In the Batak community, land tenure is not as established as in the Ata community. What consequences this will have for the Batak remain to be seen.

On the secondary resource use level, the boundary in most cases does not appear to be rigid. In fact, generalized exchange takes place at this level. For instance, between Batak and lowland migrant buyers, generalized exchange facilitated by special ties such as friendship or ceremony involves secondary resources. Among Ata and Cebuano subsistence farmers, a similar pattern exists where services, tools, and goods are exchanged.

Summary

Erosion is increasing in a number of sociocultural markers among the different ethnic groups involved in the use of resources found in the Batak and Ata communities. Noticeable among these is language. Among the Batak, a pan-linguistic orientation is developing (Tagalog becoming the *lingua franca*), although multilingualism is still prevalent. Among the Ata, their own native tongue has been completely replaced by the Cebuano language.

Present data suggest that various social processes (e.g., intermarriage, ceremonial linkages, acculturation) eliminate or alter interethnic biophysical and sociocultural markers. Through continuation of these processes over time, dissolution of ethnic differences may occur, resulting in a new concept of ethnicity or making the concept both theoretically and practically meaningless.

This study clearly suggests that ethnic differentiation is a matter of degree. When do we consider biophysical, sociocultural, and linguistic differences as markers sufficient to distinguish one ethnic group from another? When does an ethnic group begin or cease to exist? While we seem to put premium in our research concerns in ethnicity on differentiation, there seems to be another direction that we should take into account as well, i.e., homogenization of formerly heterogeneous ethnic groups. It is increasingly apparent that for practical and analytical purposes, ethnic groups are better distinguished in terms of resource use patterns rather than by traditional concepts of ethnic identity, especially in the face of growing cultural and linguistic homogenization. This study presents an analysis of ethnic groups in these terms.

The condition of the resource base has a profound effect on the subsistence strategy of the users. For instance, the disappearance of the Ata forest base has resulted in a concentration of effort on their farms by the Ata. This eliminates the energetic cost of hunting and collecting among the Ata and increases the energetic investment in farming.

The condition of the resource base can be affected by different types of activities. In the Ata area, the forest resource base for hunting and foraging was destroyed by logging. The Batak forest base remains, but increasing levels of human activity in the forest and the use of more efficient hunting tools have systematically altered the distribution and density of forest resources. Consequently, the failure rate for traditional hunting techniques is increasing. Human actions alter the distribution of resources, to which humans (not necessarily the same ones) must then adapt, sometimes resulting in a radical alteration of subsistence style, as in the case of the Ata.

In general, the Batak and lowland migrants have maintained a relatively clear demarcation of resource use up to this time, especially of forest resources, while Ata boundaries with lowlanders are extremely fluid. The exception to this is when, through the intervention of institutions outside of the community, legal title to land is held by the Ata. Relatively rigid or clearly demarcated boundaries may occur through differences in knowledge and skills, and rigid boundaries may be made fluid through social mechanisms such as intermarriage, which occurs at a high rate between subsistence farming Cebuano and the Ata.

Both the Ata and the Batak suffer from two sources of resource fluctuation: progressive and seasonal. In adjusting to these fluctuations, different resource users may overlap at some points and times and differentiate at others. Thus, rigidity and fluidity are not fixed for each type of resource.

Different levels of resource use, timing or scheduling, and the particular households involved can all affect the types of energy use relationships among ethnic groups. All or some types of resource use relationships may be pursued by a particular group of resource users. Thus, on the basis of resource use, it is difficult to fix human relationships in biological terms such as competitive exclusion, symbiosis, parasitism, or complete mutualism. Unlike most lower forms of animals, the human population exhibits a more complex pattern of resource use. For instance, among the Ata and the Batak, rigidity and fluidity of resource-use boundaries with other ethnic groups can occur at the same time. A rigid definition may be applied in the use of primary or basic resources, but a fluid one may be applied for the use of secondary resources obtained from the primary ones, or vice versa. Aside from types of resources used, it is also apparent from the Ata data that affinal ties and other ceremonial links may affect the way resource use boundaries are defined.

One final point concerns the political implications of ethnic differentiation. The effects of national and worldwide politics and economics are implicit in the Batak and Ata data, and the situation regarding control of resources must be considered in this light. For instance, changes toward fluid boundaries and generalized exchange in the Ata community are adaptations by uplanders to the immigration of lowlanders with economic and political connections beyond those available to the uplanders.

Note

1. There is still disagreement as to whether or not the Batak should be considered members of the Negrito group in the Philippines (see R. Cadeliña 1982).

References

Abruzzi, W.S.
       1981 Ecological succession and Mormon colonization in the Little Colorado River Basin. Ph.D. dissertation, State University of New York at Binghamton, New York.
       1982 Ecological theory and ethnic differentiation among human populations. *Current Anthropology* 23(1):13–35.

Barth, F.
    1969 Introduction. In *Ethnic Groups and Boundaries: The Social Organization of Culture Difference*, edited by F. Barth. Boston: Little, Brown and Company. Pp. 9–38.

Blau, P.
    1967 *Exchange and Power in Social Life*. New York: John Wiley and Sons.

Cadeliña, F.
    1983 Medical system: A vehicle for Negrito-Cebuano social interaction. *Silliman Journal* 30(3–4):157–162.

Cadeliña, R.
    1982 Batak interhousehold food sharing: A systematic analysis of food management of marginal agriculturalists in the Philippines. Ph.D. dissertation, University of Hawaii, Honolulu.

Colson, E.
    1979 In good years and bad: Food strategies for self-reliant societies. *Journal of Anthropological Research* 35:18–29.

Despres, L.S.
    1975 *Ethnicity and Resource Competition in Plural Societies*. The Hague: Mouton.

Dirks, R.
    1980 Social responses during severe food shortages and famine. *Current Anthropology* 21(1):21–33.

Endicott, K.
    1977 Some features of the economy of the Batak Negritos of Peninsular Malaysia. Paper presented at the 48th Australian-New Zealand Association for the Advancement of Science Congress, Melbourne, September 1.

Gause, G.F.
    1934 *The Struggle for Existence*. Baltimore: William and Wilkins.

Hardin, G.
    1960 The competitive exclusion principle. *Science* 131:1292–1297.

Harris, M.
  1964 *Patterns of Race in the Americas.* New York: Walker.
  1977 *Cannibals and Kings: The Origins of Culture.* New York: Random House.

Jochim, M.A.
  1981 *Strategies for Survival: Cultural Behavior in an Ecological Context.* New York: Academic Press.

Laughlin, C.D.
  1974 Deprivation and reciprocity. *Khan* 9:380–396.

Leigh, E.G.
  1965 On the relation between productivity, biomass, diversity, and stability of a community. *Proc. National Academy of Sciences* 53:777–783.

Levins, R.
  1968 *Evolution in Changing Environments.* Princeton: Princeton University Press.

Merecido, A.F.
  1983 Social vehicles and consequences of inter-ethnic flow of material culture: The case of the Cangguhub Negritos, Mabinay, Negros Oriental. *Silliman Journal* 30(3–4):127–156.

Oracion, E.
  1983a Ethnicity, intermarriage and change in the biosocial structure of the contemporary Negrito population in southern Negros, Philippines. *Silliman Journal* 30(3–4):98–110.
  1983b Negrito subsistence strategies in a changing upland ecosystem in southern Negros, Philippines. *Silliman Journal* 30(3–4):116–126.

Peterson, J.T.
  1978 *The Ecology of Social Boundaries: Agta Foragers of the Philippines.* Urbana: University of Illinois Press.

Steward, J.H.
  1955 *Theory of Culture Change: A Methodology of Multilinear Evolution.* Urbana: University of Illinois Press.

Wiens, J.
  1976 Population responses to patchy environments. *Annual Review of Ecology and Systematics* 7:81–120.

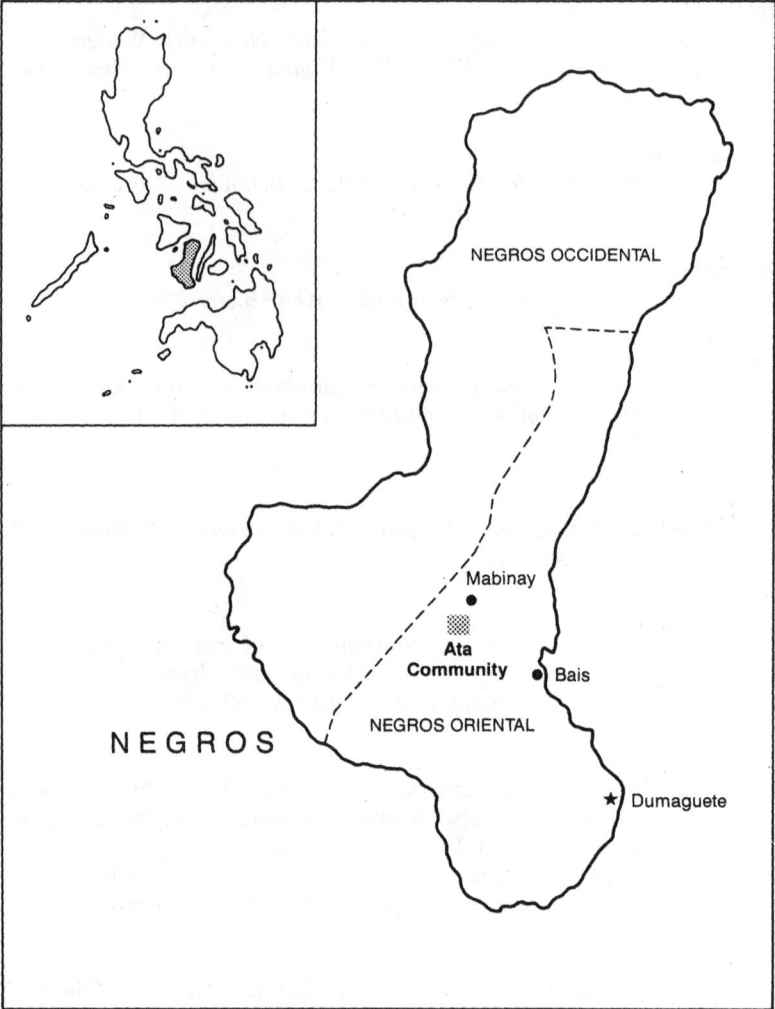

Map 4.1
Location of Ata community

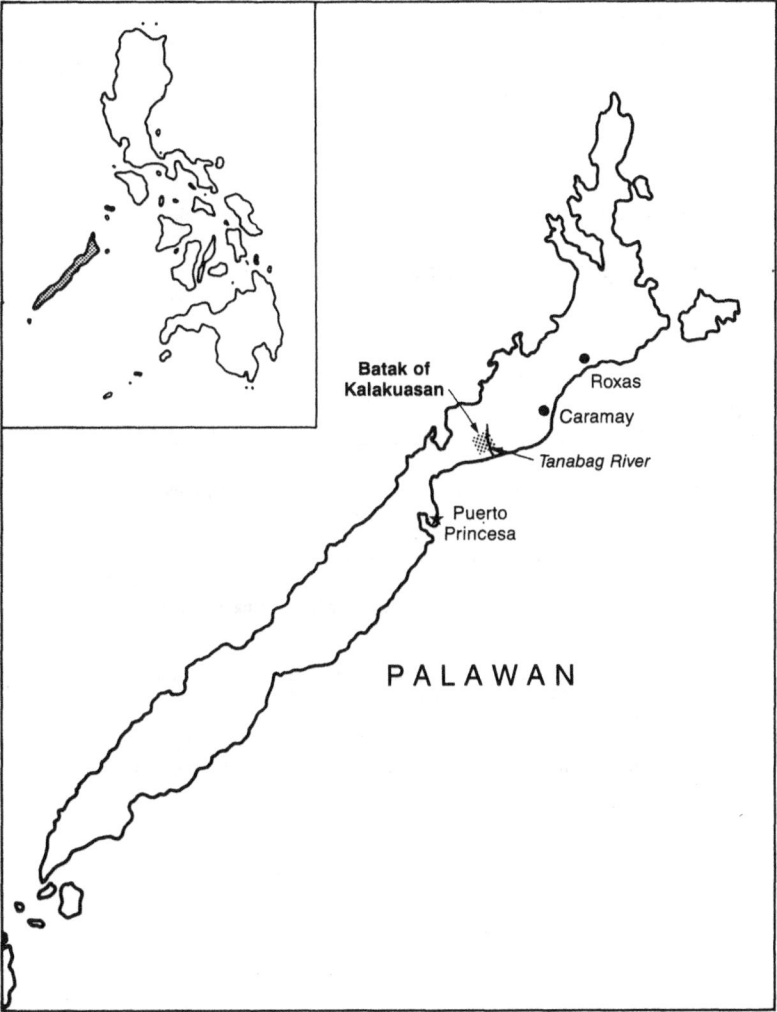

Map 4.2
Location of Kalakuasan Batak

Table 4.1

Proportion of carbohydrate food taken from various sources in one agricultural year by the Batak and the Ata

Carbohydrate food source	Batak (%)[1]	Ata (%)[2]
Domestically produced cereal	50 [3]	60 [4]
Bought or exchanged cereal	30	25
Domesticated root crop/bananas	10	15
Wild root crop	10	0
All sources	100	100

Note: For the Ata, the data were collected monthly for a one-week contact; for the Batak, the data were collected bi-monthly for a one-week contact.

[1]Seven households were observed. Data were recorded during the 1980–81 fieldwork.

[2]Four households were observed. Data were recorded during the 1983–84 survey of Oracion (1983a).

[3]Cereal largely consists of rice. Corn may be grown at the edge of the farm, but these are eaten as young corn-on-the-cob rather than as corn grits.

[4]Cereal largely consists of corn. During my 1968–69 fieldwork, few Ata households planted rice, and rice is no longer planted in the community. "Rice no longer grows," they reported.

Table 4.2
Protein sources for the Batak and the Ata

Protein sources	Batak (%)[1]	Ata (%)[2]
Hunted from forest	80	0
Fished from river and sea	10	0
Domesticated livestock	5	20
Bought or exchanged		
Salted fish	2	40
Dried fish	3	35
Meat	0	5
All sources	100	100

Note: For the Ata, the data were collected monthly for a one-week contact; for the Batak, the data were collected bi-monthly for a one-week contact.

[1] Seven households were observed. Data were recorded during the 1980–81 fieldwork.

[2] Four households were observed. Data were recorded during the 1983–84 survey of Oracion (1983a).

CHAPTER 5

ETHNIC GROUPS AND THE
CONTROL OF NATURAL RESOURCES
IN KELANTAN, MALAYSIA

Robert L. Winzeler

In this chapter I discuss an ecological approach to ethnic groups and the control of natural resources in the Kelantan plain where I have been doing research intermittently over the past eighteen years.[1] Basically I argue that although the approaches of Fredrik Barth (1969) and Edmund Leach (1954) are useful in analyzing certain basic dimensions of ethnicity, they are subject to important limitations, particularly about present-day situations.

Kelantan is the northeastern-most state of Malaysia (see Map 5.1). In the provinces of Thailand adjacent to Kelantan's northwestern boundary, the population is predominantly Malay. According to the usual economic criteria, Kelantan is among the poorest and least developed of Malaysia's states, though it is presently better off than the adjacent areas of Thailand. Consequently, a substantial southward flow of population has occurred in recent years.

For our purposes, five broad ecological zones within the state are distinguished: (1) the coastal fringe, a picturesque but impoverished strip of Malay fishing communities; (2) the Kelantan plain, extending inland along the Kelantan River for up to 40 miles, one of the largest and most densely populated rice-growing areas in the country; (3) the fringe interior, a much more sparsely settled hilly zone of both large and small rubber and oil palm holdings, combined with peasant subsistence farms; (4) the interior frontier, an expanding zone of road building, logging, land development schemes, and population movement; (5) the interior forest, a still large but rapidly shrinking zone of mountains and hills covered by tropical forest and occupied by aboriginal groups practicing swidden cultivation and engaged in hunting, gathering, and collecting, now subject to rapid infringement along the frontier.

Of these zones, I shall be mainly concerned with the area around Pasir Mas on the west side of the river and the coastal plain. Land is by far the most important natural resource within this area. Water, also a vital natural resource for agriculture and other purposes, comes in several forms. Only one, irrigation water, is controlled and provided by the government for rice cultivation, which is significant mainly in relation to it. Thus, the control of natural resources in the plain is almost entirely a matter of land control. In other zones, however, the range of important natural resources is much greater and, in addition to land, includes minerals, timber, fish and animals, and forest products (especially rattan).

Some 92 percent of the population of Kelantan is Malay, nearly all of indigenous rather than Indonesian origin. The remainder consists mainly of Chinese, Thais, and *Orang Asli* or aborigines. Overall these non-Malay groups tend to be adapted to either an urban or a rural setting, rather than both.

The Chinese, who form two distinct ethnic sectors, are an exception. The first of these are the rural Hokkien of the Kelantan plain who have been in the state for many generations and who have undergone much acculturation. The Hokkien mainly occupy a thin strip of land along both sides of the Kelantan River. Smaller enclaves of rural Chinese reside elsewhere in the plain, mostly around Wakaf Bharu; also, Chinese farmers live in a famous, old settlement in interior Kelantan at Pulai (Carstens 1980). The "urban" Chinese, on the other hand, are more recent immigrants to the state. They have undergone far less acculturation but have most of the Chinese characteristics residing elsewhere in Malaya. The Thai are also rural. Like the village Chinese, they have been in Kelantan for many generations that they have no idea of where they originally came from—they only know that it was from the north. The Thai have also undergone considerable acculturation. Most of their communities are in the northeastern corner of the plain, especially in the Tumpat district.

The Ecological Approach

In anthropology the ecological approach to ethnicity has been developed most fully and explicitly by Barth (1956, 1969), although it owes a good deal to Leach (1954) as well. In brief, Barth argues that existing concepts of the nature of ethnic groups as cultural entities are inadequate. He proposes instead that such groups should be seen as organizational entities that exist not in isolation, but in relation to others. Barth (1956:1079) suggests that ethnic groups can be compared to animal species, which adapt to various niches in the environment, with the crucial difference that members of ethnic groups can and do become members of other ethnic groups whereas such change is not possible in the realm of biological species. This tendency to alter ethnic identities is given much emphasis by Barth and by Leach.

In pursuing the species analogy, Barth argues that ethnic groups in contact will relate to one another in one (or in some combination) of four different ways: (1) They may occupy distinct niches in the environment and therefore not be in direct competition for resources; (2) They may adapt to separate territories in similar ways in which they articulate along a frontier, which can remain stable until population pressure or some other factor disturbs the balance; (3) They may occupy the same niche within the same area, which is an inherently unstable situation that could be altered in a variety of ways; (4) They could develop a symbiotic relationship involving the exchange of important goods and services, as, for example, in trading systems or in a complex mode of production (Barth 1969:19–20).

In contrast to much previous thought concerning ethnic groups, Barth argues further that an adequate understanding must account for both persistence and change. Whereas previous formulations had tended to be concerned with *either* change (process of assimilation, acculturation) or persistence (e.g., Furnival's notion of the plural society), Barth maintains that persistence and change are closely linked and explicable within the framework of his ecological approach. That is, situations are commonly found in which ethnic groups persist as organizational entities while undergoing changes in culture, membership, and modes of adaptation. For example, population growth may upset the adaptive balance in a region, compelling members of one group to expand its territory and alter its adaptation, thereby leading to the displacement of another group. Or such growth may lead some members to alter their ethnic identities and become a part of another group (Barth 1969:21–25). Barth presents several examples of such patterns, that of the Pathans of Pakistan and Afghanistan being particularly impressive. In brief, the population growth of Pathans has led to movement into new regions. The consequences, however, have varied depending upon the particular ecological circumstances. In some cases Pathans had dominated and absorbed other groups, while in others they have abandoned their identity as Pathans and assumed that of the other group (Barth 1956, 1969:117–134). Leach's (1965) discussion of Kachins and Shans of Northern Burma provides another example of such patterns of change. Both Barth and Leach present such changes as deliberate, conscious processes.

In explaining how and why shifts in ethnic identity occur, Barth argues from the perspective of "cultural ecology" and not merely "ecology." That is, when such shifts happen, it is not only because some environmental change has occurred but rather because such change is culturally significant. Individuals are motivated by the central value orientations of their culture. It is the inability to achieve or perform adequately according to these values that leads them to seek an alternative ethnic status in which it is possible to succeed or at least not fail.

For the Pathans, the central values relate to hospitality, the seclusion of women, and the council of men. Adequate performance in each of these realms requires a minimum level of wealth; that is, to be a host one must be able to provide food and lodging, to seclude women one must be able to do without their labor in the fields. So when a Pathan finds himself in an environmental situation that lacks the resources enabling him to consummate these central values, he becomes a member of an alternate ethnic sector with less demanding (or at least less expensive) standards (Barth 1969:124–134). Leach (1954:2) provides similar examples, including movement into an ethnic group with higher, more expensive standards of performance, as when a wealthy, upwardly mobile tribal Kachin chief decides to become a Shan.

Finally, such an explanation of ethnic adaptation and ethnic change, though ecological, can be distinguished from the more purely materialist approach of Harris (1979:46–76), for example. Harris assumes that the economic or adaptive base of a society determines its cultural values and institutions. Barth argues against such an interpretation in the following passage:

> ecologic feasibility and fitness in relation to the natural environ-
> ment matter only in so far as they set a limit in terms of sheer
> physical survival, which is very rarely approached by ethnic
> groups. What matters is how well the others, with whom one
> interacts and to whom one is compared, manage to perform, and
> what alternative identities are available for the individual.
> (1969:25)

This in turn repeats Leach's simpler observation made earlier concerning Upper Burma that "the ecological situation is a limiting factor, not a determinant of the social order" (1954:28). Here, at least in regard to Barth's formulation, he is not concerned with the question of where the central values of a culture initially come from. He simply accepts them as given.

I will now leave this issue aside and turn to the broader significance of the ecological approach to ethnicity. First, such analyses yield the most striking results when applied to ethnic situations set in areas close to sharply differentiated environments. Areas with abrupt changes in altitude, slope, and rainfall tend to set sharp limits on various forms of agriculture or other modes of adaptation. Such severe gradients are all characteristic of the regions from which Barth draws his examples, especially Northern Pakistan. They also seem characteristic of Upper Burma. But in most of monsoon Southeast Asia, such environmental differentiation is more moderate. Much has been made of the ecological distinctions between the uplands and the lowlands, between the hills and the plains, between inner and outer Indonesia. But these are clearly not distinctions which, when compared to Northern Pakistan, are either abrupt or based upon limiting conditions such as sharp moisture gradients. Wet rice cultivation is much more common in lowland regions because it is generally easier and not because it is impossible on mountainsides, as the terracing found in areas like Bali shows. Further, as Lehman (1967) has shown for Burma, generalizations about the ecological distribution of ethnic types have been overdrawn.

A second point needs to be considered regarding a "simple" ecological approach to ethnicity. The most striking examples occur not only in regions of abrupt environmental change but in areas where overarching political control is either weak or absent (i.e., where states are small and unable to exert much power or nonexistent). The political systems of the Northwest Frontier in Pakistan and in Burma are similarly described by Barth and Leach in terms of "organized anarchy." In such situations, ethnic processes of differentiation, accommodation, and assimilation are played out in accordance with simple, direct, local ecological and political constraints and incentives. In Southeast Asia such conditions were characteristic of much, if not all, of the region in earlier centuries (Winzeler 1976), and they may still prevail in some parts. But, in general, the current states, which were created under colonial rule, are much more powerful than those of earlier periods. The national state in Malaysia clearly has considerable power to impose its will upon its various ethnic communities and local regions. Although past relationships between local ethnic communities may have been based on the sort of ecological constraints Barth and Leach describe, these relationships are now also strongly affected by state and national policies and laws. Among other things, such policies and laws affect the control of natural resources by ethnic groups.

The Ethnic Situation in the Kelantan Plain

It may now be useful to provide a more specific analysis of the Kelantan plain that I hope fairly reflects both the utility and the limits of such analysis. As mentioned earlier, the ecological diversity of the plain is generally limited. At the most, one would be able to argue for the existence of a limited number of subniches involving somewhat different adaptations. This means that in Barth's four possible types of ethnic interaction, there have been (1) some limited tendencies to develop and exploit separate subniches in a common region; (2) some potential for conflict and competition based upon the occupation of more or less separate but adjacent or similar territories; and (3) some limited possibilities for the development of symbiotic relationships based upon the exchange of goods and services. If we move from the plain to a more complete consideration of the state as a whole, we would find a broader range of situations, though for the most part I will confine myself to discussion of the plain. Except for in the market towns, the population of the plain is overwhelmingly Malay.

The rural non-Malays—the Chinese and the Thais—reside in small enclaves; however, *where* these enclaves are located and *why* is of special interest.

The Rural Chinese

Most of the rural Chinese dwell in villages that stretch along both sides of the lower Kelantan River for much of its course through the plain. Some Malay communities are also situated along the river. Some Chinese houses are located away from the river, but the majority of the Chinese occupy a strip of land extending back only a few hundred yards or less from the river's edge. Many Chinese temples and traditional Chinese houses are situated on the river banks, facing the river. The houses, raised on low piles, are built on the river banks and modeled in a Chinese rather than a Malay style. Why are the Chinese there? There is little in the present-day adaptive patterns on which to base an answer. These Chinese are neither fishermen nor exploiters of other riverine resources. They are farmers who grow crops that could usually be grown with equal success elsewhere. Therefore, they are there clearly for reasons deriving from past, not present or even recent, conditions. The Chinese were unable to provide much help with questions about why they were where they were—questions I asked frequently in my research.

In any case, an answer to the question of why the Chinese are there involved two questions: Why were they able to be there? and Why did they want to be? Local opinion holds that the Malay rulers in the past granted them permission (or perhaps encouraged them) to settle there. We may at least infer that the Malay rulers were not opposed to such settlement for the Chinese, who certainly never had the power to impose themselves where they were not wanted. This suggests either that no one else wanted to occupy the land where they settled or that the Chinese were viewed positively as contributing in some way.

The importance of either of these considerations is difficult to understand, but I presume that both were involved. In the first place, it may be inferred that the Chinese arrived before population pressure built up in the plain. Furthermore, I presume the Malays had chosen not to occupy the strip of land immediately along the river because the soils were too sandy and porous for wet rice cultivation; therefore, they preferred to locate their villages farther back, closer to their rice fields. In the second place, the Chinese may have been considered desirable occupants from the perspective of the Malay rulers (and perhaps from that of the Malays generally) because they were involved in trading of revenue-generating areca nut and copra. This also provides the most likely answer to why the Chinese found the riverside a desirable location. That is, although the banks are not suitable for rice cultivation, they are suitable for coconut and areca nut palms and for vegetable and fruit crops. Some of them are also grown on the lower flood lands when the river drops during the dry season. The river's edge was also a good location for trading dried areca nut and copra, because such commodities were shipped by river after havesting and local processing.

The Thai

The adaptive patterns of the Thai are also interesting. Like the rural Chinese, the origin of the Thai is obscure because of their extended presence in the state. There may well have been several separate movements to the different areas of the plain where their villages are now found. Today the majority of Thais live in large villages in the northeastern part of the plain in Tumpat and Pasir Mas districts. Unlike some of the Thai of the West Coast, these Thai did not convert to Islam, unless they also simply fully assimilated into Malay society after doing so.

Like the rural Chinese, the matter of where and why the Thai chose to locate their villages is germane. There is little reason to believe that they entered the plain as conquerors or as colonists of the Siamese state to the north. Rather, they probably settled with the permission or encouragement of the Malay ruler, arriving at a time when land suitable for rice cultivation was still abundant. In some instances, they may have deliberately chosen to locate their villages closer to Malay villages to establish an interesting, symbiotic relationship.[2] However, most of them chose a nearly opposite course of settling in more remote locations, which, nonetheless, were still within the fertile alluvial plain and hence suitable for rice cultivation. In any case, they did not establish villages along the Kelantan River as did the Chinese (unless it was to abandon them subsequently without a trace), which is interesting in the light of a long, important relationship between the Thai and Chinese communities. However, unlike the Chinese who were traders, the Thai were wet rice cultivators.

The adaptive situations of the Thai and the rural Chinese clearly show how other ethnic groups form significant aspects of the environment. In settling along the river, the Chinese chose to establish themselves closer to the Malays with whom they initially established trade relations. In settling in the more remote corners of the plain, the Thai appear to have mainly chosen an opposite course. However, either initially or subsequently, they developed symbiotic relations with both the Malays and the Chinese. These relationships involved an interesting complex of ritual activities, taboos, and economic considerations, especially those involving pigs.

Pigs, which are raised to eat and to sell, form an important part of Thai subsistence. Free-roaming pigs are everywhere in Thai villages and fit well with other elements in the Thai farm operation, materially if not aesthetically. Food, such as rice bran and garbage, is readily available. Pigs foul house yards with their feces and destroy vegetable gardens, but the Thais readily accept these as trade-offs. On the other hand, the Malays are Moslems and abhor pork and pigs. The fact that Malays will not tolerate pigs around their own houses and gardens may partly explain why Thai and Malay houses are usually not located near one another.

The situation of the rural Chinese is an interesting contrast. They are fond of pork, which is an important ingredient in festival foods and ritual offerings. However, they raise pigs much less frequently than the Thais, partly because their houses are often closer to the Malays and also because pigs do not fit in as well with the other elements of rural Chinese ecology.

The Thai relationship with the Malays involving pigs had a positive and a negative side in that there are wild as well as domestic pigs in the region. For the Malays who live in areas adjacent to forest or swampland, wild pigs are a menace to crops. But for the Thai, who are avid pig hunters, they are an important resource. This means that as hunters of pigs the Thai are useful neighbors.

The Thai also provide services for other ethnic communities. They have developed a reputation for curing and dealing with sorcery. Individual Thai curers are sought out by people locally and from other parts of the country. However, the most important services the Thai provide relate to Buddhism. The Thais are devout Buddhists and support the twenty-one monasteries in the state, as do the Chinese, including the wealthier town Chinese who often give great sums of money for building projects. The Chinese worship at the temples and obtain astrological advice and amulets from the abbots, as do the Thai, though they rarely enter the monastic order. The Thai also provide funeral services for the Chinese.

Modern Developments

In the previous examples, I have discussed developmental aspects of the two principal ethnic minorities from an adaptive perspective. However, at least recently, developments in the region suggest that ethnic patterns of adaptation are more subject to political, ideological, and legal constraints than they are to the relatively simple ecological circumstances emphasized by Barth and Leach. Or to put it differently, to a large and increasing extent, ethnic groups within the plain, and probably within the state as a whole, compete for resources within a common territory.

Ethnic complementarity, specialization, and exchange continue to exist in some instances, as we have seen. But overall these are, or are becoming, secondary and perhaps marginal. The group that now dominates the more favorable sectors of the economy or environment does so not because it is unchallenged by others, but because it has the required political and legal or economic means to do so. Government positions are occupied by Malays not because members of other ethnic groups have no real interest in obtaining such positions, but because Malays have the political and legal power to exclude other groups. Similarly, the urban Chinese domination of the more profitable sectors of commerce is now aggressively challenged by Malays. This urban Chinese domination continues to the extent that it does because Malays lack serious interest in the higher, more complex realms of trade.

The changes now occurring in the Kelantan plain and elsewhere in the country broadly reflect the development of first a colonial and then a post-colonial political economy. More specifically, these changes reflect the emergence of a state that is able and inclined to exert much more authority than previously. On the control of natural resources by ethnic groups in the Kelantan plain, the provisions and policies of the Malay Reserve laws are extremely important. Under colonial rule, laws were passed in Kelantan, as in other states in Malaya, that restricted or prohibited non-Malays from ownership of land in certain regions (see Map 5.2). Such reserves were created to protect Malay control of various, mainly rural, areas within the broader context of the colonial extractive and plantation economy to which the Malays were generally peripheral. While Kelantan experienced far less colonial economic development and hence far less immigration of non-Malay peoples, the state government (which continued to exist as one of the unfederated Malayan states) enacted laws in the early 1930s that designated most of the land in the state as Malay Reserve. The only areas now excluded are a part of the town area of the capital at Kota Bharu, portions of several of the fringe interior districts, and a fairly sizable block of land in the interior. It is important to note that land already owned by non-Malays was not excluded, except in these few areas. Hence, most or all the land in the plain owned by rural Chinese and Thai is now within the Malay Reserve. This does not mean that non-Malays who owned land in the Malay Reserve had to vacate or give up ownership—rather, they had to comply with restrictions on selling and transferring of such land.

Actually, land ownership in Kelantan both in and out of the Malay Reserve is subject to restrictions concerning not only ethnic status but also native status. The latter, which concerns Malays and non-Malays, refers to a distinction between people designated as natives of the state (*anak Kelantan*), and those who are not. A native is defined as a person born in the same state as his father. Thus, in regard to land ownership, people are categorized as (1) native Malays, (2) nonnative Malays, (3) non-Malay natives, and (4) non-Malay nonnatives. Of these four types, native Malays have maximum rights of ownership regarding Malay Reserve land, and the non-Malay nonnatives have the minimum.

There are further complications. Transfers of land within the Malay Reserve are restricted by the ethnic-native status of both parties involved, for example, the owner and the buyer (i.e., Malay Reserve land owned by a non-Malay is treated differently from land owned by a Malay). In addition, a distinction is made between land transferred or inherited within a family and land to be sold. Finally, approval by one of several levels of authority may be granted, depending on the kind of transaction involved. Routine transactions are handled at the district land office, while those subject to the greatest restrictions require the approval of the Executive Council (EXCO) of the state, which includes, among others, the Chief Minister and the Sultan.

From the perspective of government laws and policies, transfers of land within the Malay Reserve involving a non-Malay are implemented as follows: (1) Transfer of ownership from a non-Malay native to a non-Malay native can and will be approved at the local level, subject to routine verification. (2) A transfer from a non-Malay to a non-Malay nonnative requires EXCO approval. Although land can only be leased for sixty-six years, a transfer can normally be obtained. (3) A transfer from a native Malay to a non-Malay native requires EXCO approval, which, however, is not now given.

The foregoing summary of government rules and policies is not all that needs to be considered to understand the relationship of ethnic status and land ownership. The way in which matters are viewed from the perspective of the ethnic groups involved must also be considered. In this regard, both class and ethnic status are relevant. Poorer, lower class individuals are less inclined to approach government offices or press their concerns on officials than are wealthier, higher class persons, irrespective of their ethnic status. The former class of people is generally shyer, more *malu*, about such things. Also, dealings with government offices and officials are more costly in several respects for poorer, lower class persons: They must pay for transportation to the office, pay petition writers to prepare written documents, and spend time traveling and waiting. Better off, higher class persons, by contrast, are relatively confident about approaching the government, are able to provide their own transportation, and can read and complete the necessary documents themselves. This is true whether the person is a Malay, a Chinese, or an Indian.

Ethnic status is also relevant, though again not so much in the case of middle-class or wealthy persons as in the case of poorer, rural persons. A Chinese businessman interested in a land deal, or in any other matter, is at least as apt as a Malay would be to pursue it. However, a poor Chinese or Thai villager would probably be less apt to do so than a Malay of similar circumstance. The fact that most land in Kelantan is Malay Reserve is usually totally irrelevant to the Malay villager because he can buy or, in most cases, sell without even being aware of such restrictions. The Chinese or Thai villager, however, is very much aware of such restrictions and believes he is under more severe restrictions than the Malay villager in buying and selling land. Although the Chinese or Thai villager has some knowledge of the restrictions associated with Malay Reserve land, he regards land transfer as difficult and arbitrary. The greater difficulty and uncertainty that a Chinese or a Thai experiences in land matters is partially due to the absence or scarcity of officials or other knowledgeable persons within his local community from whom he can seek advice. A Malay villager is likely to have a government-employed neighbor or a relative who can advise him. He also has access to the local headmen (*pengulu*). But a Thai or Chinese villager is unlikely to have such advantages. He is less likely to have access to a *pengulu* or *pengawa*, because the latter is usually a nonresident Malay.

Thus, the non-Malays appear to be losing their lands to the Malays (i.e., although a Malay native can buy land from a Chinese or a Thai, neither can buy land from the former). However, while Malay acquisition of non-Malay land may be occurring, other important factors may inhibit such a process. Unlike the Malays, rural Chinese and Thais are much more conservative about selling land. Of the three groups, the Malays are most attracted to purchasing modern consumer goods such as televisions, motorcycles, and automobiles. Also, Malays have a tendency to sell land to finance the Muslim-to-Mecca pilgrimage. And finally, the Thai and the rural Chinese are aware that once land has been sold to a Malay, it has permanently passed out of their hands.

If we examine what has actually occurred in the Kelantan plain in this regard, my information is limited and partly impressionistic. My studies show that the Thais had more land per household than did the Malays. Further, my more casual acquaintance with the plains region that contains the main Thai settlements also suggests that the Thai land base remains substantial. However, even if this observation is correct in a more long-term sense, it needs qualification by pointing out that some Thais leave Kelantan by shifting back across the border to develop landholdings in Thailand.

For the Chinese villages where I gathered systematic information on landholdings, the land base was less substantial than that of the Thai and more comparable to that of the Malays—both considered inadequate. Hence, many Chinese must seek a living elsewhere in the area or in a different region.

Land ownership in Kelantan is sufficiently complicated to make any discussion of long-term trends very hypothetical. Although land value has been greatly increasing in recent years—especially on the outskirts of towns where land is used for house sites—it is now underused for agriculture. This, in turn, is linked to the tendency of many Malay villagers to earn a living by periodic or regular wage labor elsewhere in the country or in Singapore, and the landless villagers to enter government land schemes or seek land in the state's interior.

Changes in Ethnic Identity
and the Control of Resources

Because access to land, government loans, employment, and entry into colleges and universities are restricted to ethnic status, one might expect substantial changes to occur in ethnic identity. This would also be expected based on the ecological approaches of Barth (1956) and Leach (1954), which suggest that when members of an ethnic group experience some significant changes in their adaptive pattern, they are liable to undergo a shift in ethnic identity. In Kelantan, while such a shift would provide access to previously unavailable resources, it is a difficult and, I believe, an uncommon occurrence. In effect, it would involve changing one's identity from non-Malay to Malay.

The basic problem here is that such a shift is not only a matter of self-identity and social acceptance, but also a formal, legal matter. It is true in Kelantan as elsewhere in Malaya that when one becomes a Moslem he or she is said to "become a Malay" (*masuk Melayu*); and that such a person, if believed to be sincere in the conversion and in the desire to become a part of Malay society, would be accepted, especially if married to a Malay. However such a conversion does *not* legally entitle a person to buy land or to gain access to resources reserved for Malays, for the legal definition of a Malay includes the provision that he or she be born a Malay. Moreover, such ethnic status, as well as birthplace and former name, is entered on the personal identity card that all individuals are required to have and present when transacting business with the government. Further, land transaction requires verification by local officials. Although these conditions probably do not preclude major changes in ethnic status, they do greatly restrict them.

Other shifts in ethnic identity do not face the legal obstacles of assuming a Malay identity and are therefore less problematic. These shifts, however, also have less significant implications regarding access to resources. An exception might be when a nonrural Chinese man marries a rural Chinese woman and moves into her village and assumes the identity and life-style of a rural Chinese. I have recorded a number of such instances, which reflect in minor ways the kinds of adjustments that Barth discusses. In general, however, ethnic change in the Kelantan plain tends to reflect what has already been noted—that the governmental apparatus of the modern state tends to take precedence over simpler, more traditional ecological factors.

Notes

1. My book, *Ethnic Relations in Kelantan*, Oxford University Press, 1985, contains a fuller treatment of the region and the ethnic groups discussed

in this chapter, except for the theoretical issues with which it is concerned.
2. Louis Golomb (1978) provides a detailed study of the symbiotic relationship in one Thai village in Kelantan.

References

Barth, Fredrik
1956 Ecologic relationships of ethnic groups in Swat, North Pakistan. *American Anthropologist* 58:1079–1089.

Barth, Fredrik (editor)
1969 *Ethnic Groups and Boundaries: The Social Organization of Culture Difference.* Boston: Little, Brown and Company.

Carstens, Sharon
1980 Pulai: Memories of a gold mining settlement in Ulu Kelantan. *Journal of the Malaysian Branch of the Royal Asiatic Society* 53:50–67.

Golomb, Louis
1978 *Brokers of Morality: Thai Ethnic Adaptation in a Rural Malaysian Setting.* Asian Studies at Hawaii, No. 23. Honolulu: University Press of Hawaii.

Harris, Marvin
1979 *Cultural Materialism.* New York: Random House.

Leach, Edmund
1954 *Political Systems of Highland Burma.* Cambridge: Harvard University Press.

Lehman, F.K.
1967 Ethnic categories and the theory of social systems. In *Southeast Asian Tribes, Minorities, and Nations*, edited by Peter Kunstadter. Princeton: Princeton University Press. Pp. 93–122.

Winzeler, Robert L.
1976 Ecology, culture, social organization and state formation in Southeast Asia. *Current Anthropology* 17(4):623–624.
1985 *Ethnic Relations in Kelantan.* Oxford: Oxford University Press.

Map 5.1
Kelantan and Peninsular Malaysia

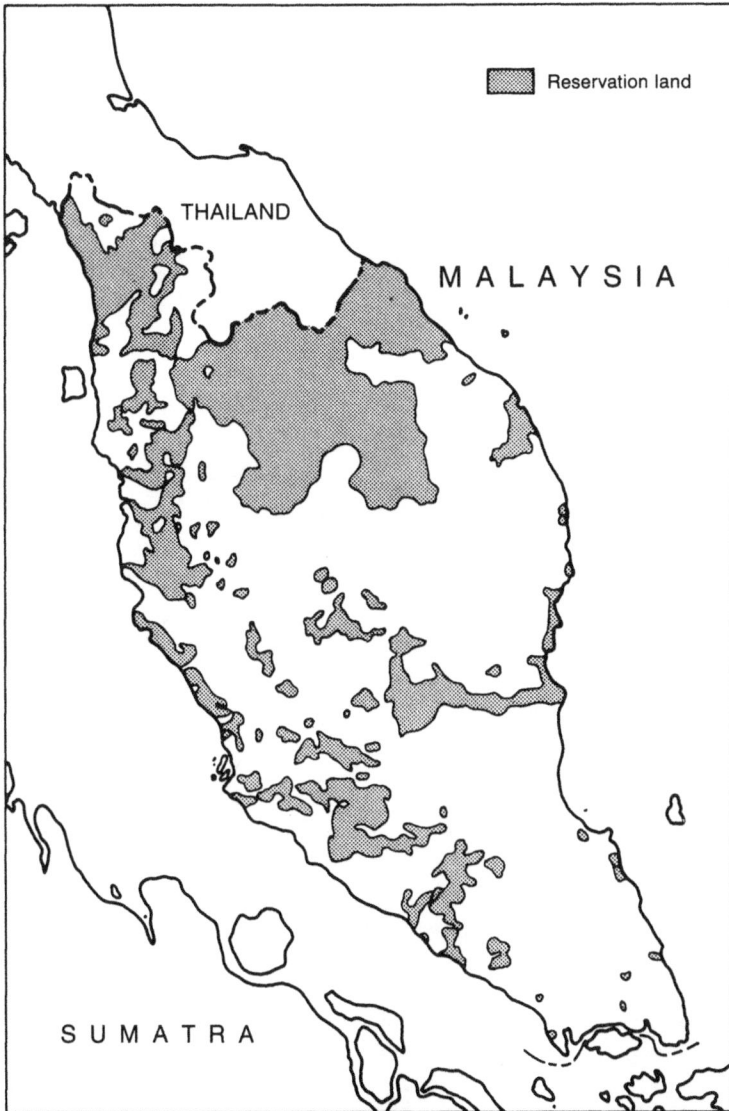

Map 5.2
Malay reservation land (shaded) in West Malaysia

CHAPTER 6

THE SEMAI:
THE MAKING OF AN
ETHNIC GROUP IN MALAYSIA

Alberto G. Gomes

One practical problem anthropologists often face is the identification and classification of the communities they research into ethnic categories or groups. Sometimes empirical evidence makes these categorizations unrealistic. A number of studies have focused on this problem of ethnicity in Southeast Asia (see Babcock 1974; King 1982; Rousseau 1975 for communities in Borneo; Dentan 1968 on Semai; Moerman 1965 on Lue; Hinton 1983 on Karen). One possible explanation for this disparity between ethnic categorization and empirical reality is that ethnic change and ethnogenesis are ongoing processes. New ethnic groups emerge (or the existing ethnic names become meaningful) and some others are accommodated (or incorporated) into larger ethnic groups.

In this chapter, I shall examine Semai ethnogenesis, beginning with a discussion of two viewpoints, that is, the outsider (government officials and anthropologists) and the insider (Semai) on Semai ethnicity. This discussion leads to a more thorough examination of the evolution of ethnic consciousness (or the emergence of ethnic "boundary maintenance") among the Semai. I shall explore factors that facilitate the occurrence and continuation of this process.

Who Are the Semai?

By official definition, Semai as a group represents a *sukubangsa* (ethnic-tribal subgrouping) of the ethnic category "Orang Asli" (aborigines). An Orang Asli is legally defined in the Aboriginal Peoples Act 134 of 1954 (amended in 1974) as:

1. Any person whose male parent is or was a member of an aboriginal ethnic group, who speaks an aboriginal language, and habitually follows an aboriginal way of life and aboriginal customs and beliefs (this category also includes a descendent through males of such persons);
2. Any person of any race adopted during infancy by aborigines, who has been brought up as an aborigine, habitually speaks an aboriginal language, habitually follows an aboriginal way of life and aboriginal customs and beliefs, and is a member of an aboriginal community;
3. The child of any union between an aboriginal female and male of another race, provided that such a child habitually speaks an aboriginal language, habitually follows an aboriginal way of life and aboriginal customs and beliefs, and remains a member of an aboriginal community.

The Jabatan Hal Elwal Orang Asli (Department of Aboriginal Affairs, hereafter referred to as JOA) follows the conventional ethnological categorization of Orang Asli into three *bangsa* (ethnic groups): Negrito, Senoi, and Melayu Asli. These categories, based on ethnolinguistic criteria, are further subdivided into nineteen *sukubangsa*, of which Semai is one. As Carey (1976:13), the first Commissioner of Orang Asli, puts it: "language, culture and physical appearance make it plain that there are, in fact, three separate ethnic groups; each group includes a number of related tribes who speak similar languages and who follow a similar way of life."

Thus, this categorization is based on the so-called primordial characteristics of the Orang Asli. In another statute, Orang Asli are classified together with Malays and natives of North Borneo (Sarawak and Sabah) to form the political category *bumiputra*, which means "son of the soil."

For administrative and census purposes, colonial officials were particularly concerned with identifying and classifying people in the colonies into ethnic groups or categories. Lacking proper comprehension of the native language, these officials often misunderstood and misinterpreted information, creating ambiguity over ethnic categorization and labels. Consequently, there is much confusion surrounding ethnic names,[1] especially as to whom or which group of people these referred to in the early literature on Orang Asli.

Skeat and Blagden's (1906) monumental work on Orang Asli, which is largely based on earlier reports on aboriginal society and culture, classifies Orang Asli into three ethnic groups: Semang, Sakai, and Jakun. In the early literature, "Sakai" was sometimes used to refer to a specific group, while other times it referred to the whole aboriginal population. This term, which implies "slave," has derogatory connotations and is held to be invidious by the aborigines. Sakai has been replaced by the label *Senoi* (which means "human" in the Senoic languages) in the specific sense and by *Orang Asli* ("original people" in Malay) in the generic sense.

Clifford (1891) was possibly the first to have defined two separate tribes of Senoi, which he labeled as Northern Senoi and Central Senoi. Later, Schebesta (1928) referred to the Northern Senoi (or Sakai) as Ple-Temiar and the Central Senoi as Semai (cf. Map 6.1). Noone (1936) claims that "Ple" is the Semang (Negrito) term for the Northern Senoi while "Temiar" or "Temer" is the Semai name for them. Noone (1936:7) further notes that Semai or "Seman" is the Temiar name "which these southern hill people have adopted from their northern neighbors."

Schebesta (1926:272) observes,

> The Semai occupy the regions of the Batang Padang, Slim, Bertam, Telom, Serau and the lesser Jelai. In physique they are certainly the purer representatives of the Sakai race. In language and culture they are clearly distinguished from the Ple-Temiar, but anthropologically there are indubitably connections between them.

Since the 1960s, four anthropologists—Robert Dentan (1965, 1968), Alan Fix (1971, 1977), Clayton Robarchek (1977), and Carole Robarchek (1980)—have published material on the Semai, but only Dentan (1975, 1976) has given some attention to the question of their ethnicity. Diffloth[2] (1968), who has carried out linguistic research on Semai language, has also raised a number of points relevant to their ethnicity.

Dentan (1968:1) says that "the word Semai refers to the aggregate of people who speak dialects of the Semai language." He considers them a "people" for the following reasons:

> First, the Semai all live in a definable geographic area. Second, they share a tradition of having been dispossessed and persecuted by non-Semai. Partially as a result, both east and west Semai tend to define their own ways of life as being not only different from but also opposite to non-Semai ways of life. Third, they have a common language which is unintelligible to non-Semai. Finally, they share a common attitude towards a great many things, most notably violence. (1968:4)

Dentan divides the Semai into "East Semai" and "West Semai" based on the extent of acculturation, but later claims that this division is an oversimplification given the political automony and heterogeneity of Semai villages. Faced with this empirical reality, Dentan (1976:74) states, "I would be hard pressed, indeed, to call it [the Semai] an ethnic group." In another paper, Dentan (1975:54) points out that "the word Semai thus does not refer to a unified set of people who are conscious of their identity as a people."

Semai "Ethnoethnicity"[3]

The Semai refer to themselves as *Senoi*, which means "man" or "human being" in their language, or by toponyms such as *mai darat* (people of the interior) or *mai cenan* (people of the hillslopes). Specifically, Semai may refer to themselves by the name of a large river valley such as *Mai Bertak, Mai Cenderi'*, or *Mai Slip*, where *Mai* means "people," or by the name of the closest town such as *Mai Tapah, Mai Kampar*, or *Mai Bidor*. Larger river valleys comprise a number of villages or *lengrii'* (territories). These *lengrii'* take the name of a smaller river or tributary. Semai then, to be even more specific, may refer to themselves and may be referred to by others familiar with Semai territorial organization by the name of their *lengrii'* such as *Mai Tenlop* (people of *Tenlop lengrii'*), *Mai Lengap* (people of *Lengap lengrii'*), or *Mai Enlak* (people of the *Enlak lengrii'*) where *Tenlop, Lengap*, and *Enlak* are also the names of rivers. The *lengrii'*, with topographical features such as rivers and ridges as boundaries, are owned by a land-owning group (*Mai pasak*, which means "true or original inhabitants") whose membership is considered by principles of cognatic kinship (Gomes 1984).

The ethnic label "Semai" is now becoming popular among the Semai people as a self-reference. This word is particularly used to refer to themselves when dealing with non-Semai. Being aware that this word is the official ethnic name for themselves, they assume that outsiders know what it means and to whom it refers. They may sometimes call themselves Orang Asli, which is the official ethnic category that includes nineteen other aboriginal groups.

In their interactions with non-Semai, Semai use the ethnic label *gob* to refer to the Malays, *gob cina* or just *cina* (a Malay term) for the Chinese, and *kling* (a Malay term) for Indians. They also use numerous nicknames when referring to these other ethnic groups. The people consider the ethnic names of "Senoi," "mai," or "Semai" as opposite to how they define a *gob*. To them, a *gob* is a Muslim and, as such, one who is circumcised and does not eat pork. The empirical fact that Semai constantly define themselves as possessing the opposite characteristics from those that apparently are central in Malay ethnic identity may have prompted Dentan (1975) to entitle his paper, "If there were no Malays, who would the Semai be?" On what basis do the people we call Semai consider themselves as forming an ethnic group?

Benjamin (1966:9) notes for the Temiar, another Senoi-speaking ethnic group,

There is little ambiguity when the term [Temiar] is used as the name of the language which the Temiar customarily speak. For all the Temiar groups that I visited, a given language is either Temiar or it is not, and this holds true despite the differences of dialect from area to area. So long as discussion is limited to linguistic matters the individual can safely declare himself to be a Temiar—by which he means, not that he was born a Temiar, but that his customary domicile is in Temiar-speaking community (and, of course, that he speaks the language fluently).

When I asked Semai why they considered other people from the neighboring villages to be Semai, some responded that these people spoke more or less the same language (*roc sama*). They, however, also pointed out that although they can understand the dialect the other Semai spoke, these people are "not really the same" (*pek sama' bernol*) as they. Nevertheless, they consider these people to be more akin to them than the Temiar (their Northern Senoi neighbors) based on linguistic affiliations. Semai from the 9th milestone village in Batang Padang will say, for instance, that people from the 25th milestone, 21st milestone, and 19th milestone all belong to the "same gang" (*gang sama'*) but are different from themselves because the others speak a distinct dialect. Similarly, Clayton Robarchek (1977:23) observes "Semai themselves are highly conscious of these differences [in dialects] and can, after hearing a visitor speak a few sentences, locate his place of origin with remarkable accuracy." The homogeneity of language or dialect among Semai *lengrii'* in geographical proximity and the relative heterogeneity and diversity of Semai dialects at greater distance imply that they have lived in dispersed clusters of villages with considerable interactions and mate exchanges among villages in a cluster and relatively little interactions between these clusters.

Before I begin to discuss the link between cultural traits and ethnicity, I wish to highlight the main points about the relationship between language and ethnicity in Semai society. First, the people recognize numerous dialects in their language and readily group people into clusters based on linguistic affiliations (Diffloth 1968:65). Second, in language, most people do not see themselves as forming a homogeneous group; in fact, they use dialectical differences to differentiate themselves from other Semai. As Dentan (1976:74) observes: "In fact, 'Semai' is primarily a linguistic category of some (but not much) significance to Semai speakers."

From my numerous conversations with Semai about their ethnicity, I discerned three cultural traits they perceive that they share, constituting a sort of "shared values system." The Semai believe their food habits, dress, and religion are different from, if not in opposition to, the set of cultural traits and values held by other ethnic groups such as the Malays and Chinese. Semai usually remark that, unlike Malays (who are Muslims), they eat pigs, monkeys, rats, and frogs. But when I point out that they are similar to Chinese in their food dietary choices, they rationalize that Chinese do not have the same food taboos (*pantak cha´nah*) as theirs. They say, "Chinese eat meat (*menhaar*) and fish (*ka´*) together in one meal, but we can't because it is *penali´*" (see Dentan 1968:Chapter 3).

In terms of dress, some Semai claim that, unlike Malays and Chinese, they are poorly dressed. A common Semai statement would be: "We are poor. We can only afford to wear old, dirty clothes." I predict that this trait will soon lose its validity in their ethnic self-identification as more and more youths don fashionable clothes (e.g., blue jeans), making them almost indistinguishable from Malay youths.

Most Semai regard their traditional religious beliefs and practices as central in defining their ethnicity. They say, "We are all Semai. We believe in *ruai* (head soul), *klook* (body soul), *gunik* (spirit familiar), *nyani* (spirits), *halaa´* (shaman), and we have *kubut* and *asik* (ritual ceremonies)." Interestingly, people who attend the ritual ceremonies are considered "alike," but the *gunik* (spirit familiars) may be ethnically different. I know of a *halaa´* who claims that he has a Malay *gunik* and an European *gunik*; in fact, during the seances he speaks in Malay when he is supposedly possessed by the Malay *gunik* and speaks Malay with an English accent when possessed by the European *gunik*. During the ritual ceremonies I attended, the *halaa´* called out to his *gunik* not to be afraid or shy of my presence and rationalized that I was just like "one of us." Dentan (1975:51) reports a similar experience.

Almost all the people I interviewed responded positively to the question of whether Semai who have become Christians are still considered Semai. Most, however, felt that Semai who became Muslims usually ceased to be regarded as Semai. They are said to have "entered Malay" (*muit gob*). Since Islam assumes a central role in Malay ethnicity, adopting this religion is usually viewed, at least in Malaysia, as "becoming Malay." Adopting Islam also means that new converts must stop certain practices, such as eating pork and other wild meats considered central to Semai cultural practices and associated with their ethnic identification. Furthermore, the tolerance shown to Muslim Semai and consequently the relative ease of intermarriage between Malays and Muslim Semai may have facilitated the integration of the Muslim Semai into the Malay community. Intermarriages between Malays and non-Muslim Semai are generally discouraged and looked upon with disdain by the Malays.

The Process of Incorporation and Semai Ethnogenesis

Early writers (Annandale and Robinson 1903; Cerruti 1908; Evans 1923; Schebesta 1928; Skeat and Blagden 1906) reported that Batang Padang Semai lived in dispersed hamlets in isolated regions of the forests. These Semai were also described as politically autonomous from Malay and colonial rule, apart from the occasional slaving raids on them by Malays (see Endicott 1983). Semai, especially those knowledgeable in native folklore, talk of warfare called *praak sangkil*[4] between their forefathers and Malays that occurred around the turn of the twentieth century. They claim that the Malays waged war because the early Semai were unwilling to comply with the Malay ruler's wishes for them (Semai) to become Muslims.[5] Politically, this war can be interpreted as a Malay attempt to incorporate the Semai into the Malay state of Perak. As I discussed earlier, the adoption of Islam by the Semai would have resulted in eventual integration of the Semai into the Malay community.

One may be tempted to argue that a sensible way for the people to counteract Malay harassment would be by their consolidating into larger groups, as did some communities in Borneo (Rousseau 1975), through forging "ethnic" solidarity. Instead, early reports and native oral histories relate that the usual response to Malay raids was flight and not fight. The question then arises as to why the people did not forge ethnic solidarity and consolidate to form large groups to counter the Malay raiders? I will suggest a number of reasons.

First, the Semai, lacking sophisticated military techniques and technology, could not match the Malays in warfare and weaponry techniques. The Malays, especially the migrants from Indonesian islands, were experienced in warfare; their frequent feuding and piracy are documented in historical records. But in the Semai folk history, the Semai are said to have emerged victorious not through military means but through magical ways. These legends describe a powerful shaman who led the Malay warriors into a swamp and then, through incantations and other rituals, directed a swarm of bees to attack and kill the Malays.

Second, the forest environment provided the Semai with ample hiding places from the Malay raiders. Unlike the Malays who generally kept away from the forests, the Semai were superbly adapted to this environment. They lived in extended-family longhouses situated on hill tops or steep slopes that provided an excellent view of the access routes to their residences. Living in small hamlets also made it easier for them to move rapidly to new locations.

The Semai, however, were eventually partially incorporated into the Malay state. But most of them remained non-Muslims, making them distinct from the Malays and preventing their total integration into the Malay community. The dispersed Semai settlements, perhaps partly from Malay incursions and the concomitant lack of ethnic solidarity among the Semai, may have gradually changed as the colonial regime consolidated its power in the Malay states, thus removing Malay threats against the Semai.

As Dodge (1981:9) wrote, "The colonial regime introduced a policy of paternalistic protection and gradual integration into the wider society" for the aborigines. The colonial regime's ban on slavery in the late nineteenth century was perhaps the most effective policy on Semai-Malay social relations. Because of this protection, the Semai became more confident and less fearful of open interactions with the Malays.

In recent years, the Semai have become increasingly ethnically conscious. Although some of them do not yet see themselves as forming an ethnic group, many do so based on previously mentioned cultural traits and the sharing of a common set of values and aspirations. A prominent Semai leader, at a meeting to decide whether Semai from Batang Padang should form an association, said,

> We need an association. We must be united. Politicians tell us that we can go to UMNO [United Malays National Organization—the principal-making political party] if we have problems. We can't do that, UMNO is for *gob* [Malays]. We are not *gob*, we are Semai.

I consider four critical factors underlying the development of ethnic consciousness and ethnic "boundary maintenance" now operating in Semai society. These are (1) changes in Semai social organization in the early twentieth century, (2) the *Emergency*, (3) Malaysian state (particularly JOA) policies and development programs, and (4) land rights and control of resources.

Although I treat these factors separately here, they are interrelated and complementary and are ordered not by importance but from the standpoint of historical sequence.

Changes in Semai Social Organization

According to Semai folk history, in the early 1900s a Semai headman, Bah Busuh from the 8th milestone region in Tapah, established himself as the leader of all Semai in the Batang Padang region. He was recognized as the Semai leader, or chief, not only by the Semai but by the Malay rulers. He received a letter of authority (*Surat Kuasa*) from the then Sultan of Perak and was invested the title of *Maharaja Lelawangsa*. Bah Busuh is attributed with having initiated two important changes in Semai organization. First, he is said to have established native courts and a system of Semai law (modeled after Malay *adat*, customary law). Court trials (called *bicara'* in Semai) were convened to punish law breakers by imposing suitable fines. *Bicara'* were also held to resolve disputes among Semai individuals, families, or even villages.

Second, Bah Busuh initiated meetings among all the regional Semai village headmen to discuss establishment of their own system of courts and laws and territorial organization. At that time, the rights to land and resources were a frequent source of dispute between the villages. These conventions resulted in an elaborate system of land ownership and rights to resources (see Gomes 1984), which, in turn, led to consolidation of the numerous dispersed Semai villages, more social linkages between these villages, and enhancement of their ethnic solidarity.

Bah Busuh is said to have been on cordial terms with the Malays, often assuming the role of spokesman for the Semai when dealing with Malays. Possibly to reinforce the cordiality between Semai and Malays, Bah Busuh presented fruits, mushrooms, rice, and forest products to the Malay rulers. Although these gifts were collected from all the villages in the region, it does not imply that the Semai were fully incorporated into the Malay state. From the Semai standpoint, the Malay rulers may act favorably toward them if the Semai are harassed by their Malay neighbors. Thus, the Semai could remain somewhat politically autonomous.

The *Emergency* (1948–60)

Most Semai and other Orang Asli were greatly affected by the Communist insurgency, officially known as the *Emergency*. The Communist guerrillas operated from jungle bases and often relied on the jungle-dwelling Orang Asli, including Semai, for food supplies and strategic information. Thus the Orang Asli were for the first time regarded as politically and strategically important, not only by the Communist insurgents but even more so by the government. Strategically, it was then important for the government security forces to resettle the Orang Asli away from the jungle to deny the Communists their support.

In the Batang Padang area, most Semai were regrouped (either at their own free will or forcibly) into three settlements, which were guarded by security forces. As these camps had poor facilities and were overcrowded, numerous Semai residents contracted diseases and many died as a result. This strategy was eventually abandoned following criticisms of the government's bad treatment of the aborigines (Short 1975).

In place of resettlement, the government set up jungle forts manned by security forces and aimed to keep track of the deep-forest aborigines. These forts were set up in key areas where numerous Orang Asli were suspected of supporting the Communist terrorists. The Orang Asli were encouraged to stay near these forts and to make use of the government's security and medical services.

The Communist insurgents set up an organization called "Asal," which means "aboriginal," aimed at enlisting aborigines to provide armed help. An Asal group was organized for each river valley and was usually led by a prominent Orang Asli headman or chief.

In response to the constant harassment from the security forces and the insurgents, the Orang Asli in certain areas met and agreed to a pact of mutual help called "river valley pacts" (Short 1975). The strategy was that one village would support the Communists and another, the government. Neither would give any information to either side in case it might directly or indirectly endanger the aborigines. All the other villages would remain neutral. This pact ensured the aborigines that whichever faction emerged victorious at the end, they would not be adversely affected; they would have at least one of them with the victors to support the others.

The three strategies of resettlement by the government, the formation of Asal groups by the Communists, and the river valley pacts by the aborigines all resulted in the establishment of supravillage structures. Although these structures were temporary, they reinforced the intervillage linkages that were initially fostered by Bah Busuh's efforts. The common effect of these strategies was the grouping of people from different remote villages. In the river valley pacts, people came together to seek a solution to their common predicament. The meetings and the sharing of similar aspirations and values may have strengthened the "we" (*hii´*) feeling among them.

Malaysian State Policies and Development Programs

The Malaysian government policy is to protect and advance Orang Asli (government of Malaysia 1961). It stresses that the "advancement" policy is with the view of "ultimate integration" of Orang Asli within the Malay community. But it also states that "the aborigines, being one of the ethnic minorities of the Federation, must be allowed to benefit on an equal footing from the rights and opportunities which the law grants to the other sections of community." It thus treats the Orang Asli as one ethnic group despite the extent of diversity among the different tribes comprising the category.

The State, mainly through the JOA, has implemented many development programs for the Orang Asli. The JOA performs three main tasks of "the provision of medical treatment, education, and rural development" (Carey 1976:300). The rural development projects include (1) resettlement (or the officially more preferred term, regroupment) of Orang Asli into "pattern settlements" where they are housed in Malay-type dwellings, provided with a piped water supply, encouraged to cultivate cash crops such as rubber, oil palm, and fruit trees in specially designated plots of land, and provided with facilities such as a school, community hall, health clinic, and sanitary conveniences; (2) introduction of cash crops and fish farms to unresettled Orang Asli; and (3) training of Orang Asli in agricultural skills.

The JOA also employs many aborigines as administrators, clerks, and medical personnel; some aborigines are recruited as policemen or as soldiers in the special army force called "Senoi Praak" (which means "fighting men" in the Senoic languages). The public radio network broadcasts special programs for Orang Asli in the two major Senoi languages, Temiar or Semai.

Although the State-sponsored development projects and the educational programs have not been entirely successful (see Endicott 1979 for a discussion of Orang Asli development), they may have had similar effects on Semai social organization to the *Emergency* strategies discussed earlier. They brought Semai together in resettlement projects or in schools and consequently provided ample opportunities for the people to establish social linkages and to share their values and aspirations. Similarly, Semai working for JOA or serving the army and police could interact with other Semai and Orang Asli from other areas, while the Orang Asli radio broadcasts made them more ethnic conscious. On numerous occasions, I overheard Semai who were comparing their ritual songs (*Jenulak*) with the songs on radio sung by Semai from other areas conclude that these other people must be like them since the songs were quite similar. All this, as Carey (1976:68) observed "has undoubtedly given the aborigines a new feeling of community, of belonging to one particular ethnic group." The contacts, social linkages, and opportunities to share their experiences may have all provided the impetus for the development of Semai and Orang Asli ethnic consciousness.

Land Rights and Control of Resources

In recent years, Semai (at least those from the Batang Padang region) have become exceedingly involved in simple commodity production. They harvest fruits such as durian (*Durio* spp.) and petai (*Parkia speciosa*) and collect forest products such as rattan and bamboo to earn cash. Although they also traded such products in the past, they were not as extensively involved in the market economy as they are now.

A clear definition of rights to land and resources is crucial in commodity production. It is also important that these rights are commonly recognized and legitimized. To ensure a reproduction of this form of production, it is imperative that the Semai have legal ownership of, and control over, the land on which their fruit trees thrive. However, they do not yet have legal title to the land they claim through their traditional system of landownership. Most of the land they use is officially classified as government-owned forest reserve.

Since the early 1960s, JOA officials have requested for converting a portion of the forest reserve in the Batang Padang region into an aboriginal reserve. Only in 1983 did the Forestry Department agree to a conversion, and the District Office carried out land surveys to demarcate the areas involved. Meanwhile Semai were expressing their concern over land rights as they were becoming aware that they did not have legal title to their land. Although they were informed that they would be given rights to certain parts of the land they claim, most Semai felt that it was a trick by the *gob* (Malays) to take control of their territory. Their suspicion was strengthened after some of them who worked for the survey team realized that only a section of their *lengrii'* would be legally recognized as theirs. This would mean that they would automatically lose their rights to the fruit trees outside of this section.

. The land-rights issue, as experienced by many displaced tribal populations in the world, had the effect of bringing people together with similar aspirations. The Semai elites brought people together to discuss ways of resolving the issue. An association with a solely Semai membership was formed but is now waiting to be officially registered. Its main objective is to channel Semai grievances about land rights and economic exploitation to the proper authorities. The formation of such an ethnically based association indicates that ethnicity has become an important Semai consideration.

Conclusion

Southeast Asia is renowned for its ethnic diversity. The nation-states that were once colonies in the region embarked on a policy of ethnic integration upon gaining independence. This policy was in opposition to the colonial policy of ethnic segregation or "divide and rule." The new governments linked ethnic integration to successful economic development. One would therefore assume that the implementation of such a policy will inevitably lead to the decline or demise of ethnic groups. However, as the Semai case discussed in this chapter illustrates, the integration policy may, in reality, result in just the opposite of what has been intended.

As is common with many State policy implementations guided by the assumption that since it has worked elsewhere it has a good chance of working here, the Malaysian State policy of Orang Asli integration into the Malay community has failed. In fact, it has provided an impetus for the development of Semai ethnic consciousness and identity (a process I term "ethnogenesis"). However, the failure of the integration policy does not imply that the Semai have remained removed from the "mainstream of society." Their increasing involvement in the market economy, especially through their trading of commodities (fruit and forest products) and their purchasing of market-produced food and goods, has resulted in the incorporation of the Semai economy with the wider Malaysian and international economy.

Being drawn into the multiethnic Malaysian society where political, economic, and social relations are influenced by ethnicity and ethnic identities, the Semai and the other Orang Asli ethnic groups find it necessary to align or associate themselves with an ethnic identity. They do not want to "become Malay" (*Muit gob*)[6] because the Malay-Semai relations were and are generally not cordial, and as Dentan (1975:50) has correctly noted, "the Semai normally identify themselves as the opposite of Malays." In fact, as Dentan (1975:54) has pointed out, "What unity the Semai feel they have seems to come from their constant contrast of how they live with how Malays live." Many Semai have become Christians or Bahai and may have been more receptive to these faiths than Islam to dissociate themselves further from the Malays.

According to the outside view, Semai ethnicity is based on ethnolinguistic markers. The Semai themselves consciously manipulate these markers to situate themselves and others in a process of "exclusion and incorporation" (Barth 1969:10) into groups. In the past, Semai did not see themselves as forming an ethnic group. In fact their intervillage relations were usually limited to a cluster of villages or, at most, to a row of villages along a river valley. Instead of fostering ethnic solidarity among the Semai, the Malay incursions in Semai land and the slave raids had imposed further dispersal of the Semai villages into the fairly inaccessible forests. The protection given by the colonial state at the end of the nineteenth century allowed the Semai to consolidate into larger settlements.

The development of Semai consciousness, or ethnogenesis, was aided by a number of factors. Internal factors—such as the emergence of Semai leaders and elites and their efforts in consolidating the dispersed Semai settlements—and external factors—such as the government's resettlement programs, Communist Asal organizations, and the Semai river valley pacts—expanded intervillage contacts among the Semai. The Malaysian State policies and modernization programs created further opportunities for intervillage contacts and supravillage organizations. Ethnogenesis was further fostered by the growing awareness and response of the Semai to the threat of losing control over resources such as land and fruit trees. This brought about common interest among the Semai and triggered formation of an association using ethnicity as a basis for grouping and membership. The Semai case therefore provides empirical support to the proposition based on other studies (Despres 1975; DeVos and Romanucci-Ross 1975) that the genesis and persistence of ethnic groups are related to intergroup competition for scarce environmental resources.

Acknowledgments

Field research on the Semai was carried out between September 1982 and February 1984 and was supported by the Australian National University. In Malaysia, I wish to thank the Department of Aboriginal Affairs (JOA) for its assistance, and I offer my heartfelt gratitude to the Semai of Upper Batang Padang, particularly the people of Batu Sembilan, and especially to Bah Akeh and Bah Toneh (Anthony Williams-Hunt). In writing this chapter, I wish to thank Dr. Nicholas Peterson for his insightful suggestions and useful comments. The thoughts and ideas presented in this chapter have been derived from discussions by conference participants on "Ethnic Diversity and the Control of Natural Resources in Southeast Asia," 22–24 August 1984, held at the University of Michigan, Ann Arbor. Finally I wish to acknowledge the assistance of the East-West Center and Dr. A. Terry Rambo.

Notes

1. Early writers have accorded different ethnic labels to refer to the Semai. Blagden (1903) called them "Central Sakai," Clifford (1891) referred to them as "Senoi," and Annadale and Robinson (1903) and Cerruti (1908) labeled them *mai darat*, meaning "people of the interior" in Semai. In arguing for the preference of "Central Sakai" as an ethnic label for this group, Wilkinson (1939:21) noted that "Senoi" and *mai darat* are also used as self-reference by the "Northern Sakai," who he claims represent "a different race." The apparent confusion over the ethnic names of these groups can best be explained in the words of Williams-Hunt (1952:19): "A group may have three names, one used by itself, one by which it is called by the Malays, and one or more by which it is known to adjacent Aboriginal groups." More recently, Howell (1981 on Che Wong), Laird (pers. com. on Temoq), and Nowak (pers. com. on Hmak Betisek) have noted the ambiguity of ethnic labels for the Orang Asli groups they studied.
2. Diffloth (1968:65) noted, "Semai is a term used by the Jabatan Orang Asli, probably following H.D. Noone, to identify the Senoi aborigines living in an area roughly circumscribed by a line joining Ipoh, Teluk Anson, Tanjong Malim, Raub, Kuala Lipis and by the Kelantan-Pahang boundary to the north."
3. I follow Dentan (1976) in using this term.
4. *Praak* is derived from the Malay word *perang* for war. My informants were not clear on what *sangkil* meant.

5. Cerruti (1908:109) related, "Some centuries later, in an era of fanaticism, invasions [by Malays] were made upon them [Semai] with the object of converting them to Mohammedism (*sic*) but the only result was fire and bloodshed and after each conflict the surviving Sakais fled further into the forest (into those parts which had never been before explored) or to the natural strongholds of the far off mountains." Clayton Robarchek (1977:26) also noted that "Semai folklore describe an early war between the two groups [Malays and Semai]."

6. As Carey (1976:35) observed, "Typically, an Orang Asli thinks of himself first as a member of a particular tribe, second as a member of general aboriginal community, and third as a Malaysian" and predicts that "it is extremely unlikely that the Orang Asli will be absorbed by any of the large ethnic groups such as the Malays."

References

Annandale, Nelson, and H.C. Robinson

    1903 *Fasciculi Malayenses: Anthropological and Zoological Results of an Expedition to Perak and the Siamese Malay States, 1901–2; Anthropology, Part 1.* London: The University Press of Liverpool.

Babcock, Tim G.

    1974 Indigenous ethnicity in Sarawak. In *The Peoples of Central Borneo*, edited by J. Rousseau. *Sarawak Museum Journal*, Special Issue 22. Kuching, Malaysia. Pp. 191-202.

Barth, Fredrik (editor)

    1969 *Ethnic Groups and Boundaries: The Social Organization of Culture Difference.* Boston: Little, Brown and Company.

Benjamin, Geoffrey

    1966 Temiar social groupings. *Federation Museums Journal* 11:1-25.

    1985 In the long run: Three themes in Malayan cultural ecology. In *Cultural Values and Human Ecology in Southeast Asia*, edited by Karl L. Hutterer, A. Terry Rambo, and George Lovelace. Ann Arbor: University of Michigan Center for South and Southeast Asian Studies, Paper No. 27. Pp. 219-278.

Blagden, C.O.

    1903 The comparative philology of the Sakai and Semang dialects of the Malay peninsula. *Journal of the Royal Asiatic Society of Great Britain, Straits Branch* 39:47-63.

Carey, Iskandar
    1976 *Orang Asli: The Aboriginal Tribes of Peninsular Malaysia.* Kuala
        Lumpur: Oxford University Press.

Cerruti, G.B.
    1908 *My Friends the Savages.* Translated from Italian by I. Stone
        Sanpietro. ´Como, Italy: Tipografia Cooperativa Commense.

Clifford, Hugh
    1891 Some notes on the Sakai dialects of the Malay peninsula. *Journal
        of the Royal Asiatic Society of Great Britain, Straits Branch* 24:13–
        29.

Dentan, Robert K.
    1965 Some Senoi Semai dietary restrictions. Ph.D. dissertation, Yale
        University.
    1968 *The Semai: A Nonviolent People of Malaya.* New York: Holt,
        Rinehart and Winston.
    1975 If there were no Malays, who would the Semai be? *Contributions
        to Asian Studies* 7:50–64.
    1976 Ethnics and ethics in Southeast Asia. In *Changing Identities in
        Modern Southeast Asia*, edited by D.J. Banks. The Hague:
        Mouton. Pp. 71–81.

Despres, Leo S.
    1975 *Ethnicity and Resource Competition in Plural Societies.* The
        Hague: Mouton.

DeVos, F., and L. Romanucci-Ross
    1975 *Ethnic Identity: Cultural Continuities and Change.* Palo Alto, CA:
        Mayfield Publishing Company.

Diffloth, Gerard F.
    1968 Proto-Semai phonology. *Federation Museums Journal* 13:65–74.

Dodge, Nicholas N.
    1981 The Malay-Aborigine nexus under Malay rule. *Bijdragen Tot De
        Taal-, Land- en Volkenkunde* 137:1–16.

Endicott, Kirk M.
 1979 The impact of economic modernization on the Orang Asli (aborigines) of northern Peninsula Malaysia. In *Issues in Malaysian Development*, edited by J. Jackson and M. Rudner. Singapore: Heinemann Educational Books. Pp. 167–204.
 1983 The effects of slave raiding on the aborigines of the Malay peninsula. In *Slavery, Bondage, and Dependency in Southeast Asia*, edited by A. Reid and J. Brewster. Brisbane: University of Queensland Press. Pp. 216–245.

Evans, Ivor H.N.
 1923 *Studies in Religion, Folk-lore and Customs in British North Borneo and the Malay Peninsula.* Cambridge: Cambridge University Press.

Fix, Alan G.
 1971 Semai Senoi population structure and genetic microdifferentiation. Ph.D. dissertation, University of Michigan, Ann Arbor.
 1977 *The Demography of the Semai Senoi.* Museum of Anthropology, Anthropological Paper No. 62. Ann Arbor: University of Michigan.

Gomes, Alberto G.
 1984 Capitalism, landrights and the Orang Asli: A case study of the Upper Batang Semai of Malaysia. Paper. Government of Malaysia.
 1961 Statement of Policy Regarding the Administration of the Orang Asli.

Hinton, Peter
 1983 Do the Karen really exist? In *Highlanders of Thailand*, edited by J. McKinnon and W. Bhruksasri. Kuala Lumpur: Oxford University Press. Pp. 155–168.

Howell, Signe
 1981 The "Che Wong" revisited. *Journal of Malayan Branch of Royal Asiatic Society* 54:57–69.

King, Victor
 1982 Ethnicity in Borneo: An anthropological problem. *Southeast Asian Journal of Social Science* 10:23–43.

Moerman, Michael
 1965 Ethnic identification in a complex civilization. Who are the Lue? *American Anthropologist* 67:1215–1230.

Noone, Herbert D.
    1936 Report on the settlements and welfare of the Ple-Temiar Senoi of
    the Perak-Kelantan Watershed. *Journal of the Federated Malay
    States Museums* 19.

Robarchek, Carole
    1980 Cognatic kinship and territoriality among the Semai-Senoi.
    Federation Museums Journal 25:89–102.

Robarchek, Clayton A.
    1977 Semai nonviolence: A systems approach to understanding. Ph.D.
    dissertation, University of California, Riverside.

Rousseau, Jerome
    1975 Ethnic identity and social relations in Central Borneo. *Contributions to Asian Studies* 7:32–49.

Schebesta, Paul
    1926 Sakai in Malakka. *Archiv. F. Rassenbilder Bildaufsatz* (München)
    9:81–90.
    1928 *Orang Utan*. München: Brockhaus.

Short, Anthony
    1975 *The Communist Insurrection in Malaya 1948–1950*. London:
    Frederick Muller.

Skeat, W.W., and C.O. Blagden
    1906 *Pagan Races of the Malay Peninsula*. London: Macmillan.

Wilkinson, R.J.
    1939 Some "Sakai" problems. *Journal of the Malayan Branch of the
    Royal Asiatic Society* 17:131–133.

Williams-Hunt, P.D.R.
    1952 *An Introduction to the Malayan Aborigines*. Kuala Lumpur: The
    Government Printer.

Map 6.1
Approximate distribution of Orang Asli groups
(after Benjamin 1985)

CHAPTER 7

PLANT PRODUCTS AND ETHNICITY IN THE MARKETS OF XISHUANGBANNA, YUNNAN PROVINCE, CHINA

Pei Sheng-ji

Ancient Chinese documents have convincingly shown that, for at least the past 2,000 years, trade has played an important role in the economy of southwestern Yunnan Province. Trade among the Dai (T'ai) people, the Han Chinese, and small groups of mountain tribes has been of particular significance. The trade in Puer tea provides a good example of this phenomenon.

"Puer tea" is a commercial name for a broadleaf tea plant (*Camellia sinensis* var. *assamica*) that has long been one of the most important native products of Xishuangbanna prefecture in southwestern Yunnan. The tea plant may have originated in this part of China where it was eventually domesticated by the Xishuangbanna natives several thousand years ago. Even now in certain areas that are maintained as natural forest lands, visitors can find scattered remnants of primitive tea plantations containing tea trees hundreds of years old and roughly 10 to 20 meters tall. The contrast of these old plantations with contemporary Chinese tea plantations is very striking.

Native farmers collect young shoots with leaves from tea trees all year round and process them into crude tea for sale in local markets. In 1936 about 3 million kilograms of Puer tea were produced in Xishuangbanna[1] (Zheng 1981). Until the 1950s, Han Chinese traders came from Puer city to purchase and transport this crude tea by horsetrain back to Puer, about 150 miles away, where it was processed into "bricks" or loose tea in family workshops. The processed tea was then distributed throughout the rest of Yunnan Province, shipped in large quantities to Tibet and Sichuan, and traded to markets in Guangdong and Hong Kong. The Dai word for the tea is *yela*, but it is known as "Puer" tea, rather than *yela* or Xishuangbanna tea, in markets and teahouses throughout China.

Puer tea played an important role in the historical development of economic and political ties among the many ethnic groups of Xishuangbanna and the Han Chinese, the Bai, and the Tibetan peoples of other parts of the country. Tea transport to distant Dali district in northwest Yunnan Province and trade with the Bai people[2] had already begun 1,200 years ago during the Tang dynasty (Zheng 1981). The tea was later traded to Tibet and became an indispensable drink to the Tibetans. In exchange, the Tibetans supplied horses to ancient Xishuangbanna in what was called an "exchange of tea and horses" (*cha-ma-jiao-hua*). Subsequent to the seventeenth century, Puer tea was available throughout the country and its fame brought an onrush of merchants to Xishuangbanna.

The Qing government (A.D. 1644–1911) set up its Official Tea Bureau at Simao near Puer to control the tea trade and to collect taxes. Branch offices were opened in every tea growing area and the tea growers exchanged crude tea for salt from merchants at the rate of approximately 100 kilograms of crude tea for each kilogram of salt. Today, however, the tea and salt price ratio has been reversed and 1 kilogram of crude tea can be exchanged for roughly 10 kilograms of salt. The preliminary processing of the tea leaves has been improved, and hundreds of workshops for black tea and green tea have been established in the area during the past thirty years. The different types of tea have increased in number, and quality has improved with scientific and technical advances. "Yunnan Black," "Yunnan Green," and "Puer Tea" are products well established in the markets of modern China.

The story of Puer tea also hints at an important trade relationship among the Bai, the southwest China Tibetans, and the minority peoples of Xishuangbanna and the Han Chinese that has long been a significant, historic fact. Although some Chinese agronomists, historians, and ethnologists have recently begun to investigate certain elements of this historic trade relationship in their field studies, most aspects of this and other more contemporary inter-ethnic relationships involving human-environmental interactions have not been systematically researched, described, or documented.

An important site for research on such interethnic relationships will be the markets of Xishuangbanna, which have long played a central role in the historic interactions of the area's various ethnic groups. Many plant products from both cultivated fields and natural forests, from the lowlands and uplands, are today sold and exchanged in these markets just as they were throughout much of the historic past.

Markets are particularly ideal for ethnobotanical research. From an ethnobotanist's point of view, the market represents a place of intense interaction not only among people of different ethnic groups but also between people and plants. People require plants to fulfill certain biological, cultural, and economic needs and also frequently depend upon organized exchange structures to obtain some of the plant products they need. In addition, the nature of the market is to select certain kinds of plant products, a pattern of selection that often results in a more intense interaction between the particular plant population and the human population that supplies them for the market (Bye and Linares 1983).

In this chapter, I will focus upon the plants found in the markets of Xishuangbanna. Following brief descriptions of the area, its environment, and population, I will report on the common plant products that I have observed in local markets over the past twenty years and discuss how these products reflect the interactions among different ethnic groups in the area.

The Area, Its Environment, and People

Xishuangbanna prefecture is located south of Yunnan Province near the borders of Laos and Burma (Map 7.1). The prefecture covers 19,220 square kilometers, stretching from 21°10´ to 22°40´ north latitude and from 99°55´ to 101°50´ east longitude. Administratively, Xishuangbanna is divided into three counties—Jinghong, Mengla, and Menghai—and includes nineteen communes and three towns.

Approximately 94 percent of the total area consists of mountainous and hilly terrain, with river valleys making up the remaining 6 percent. The Lanchang River (the upper Mekong) and its branches and tributaries traverse the whole area. The general topography slopes from north to south, and most of the area is between 500 and 1,000 meters above sea level.

Xishuangbanna has a tropical monsoon and hot, humid river valley climate typical of the southern portion of Yunnan Province. Although considerable deforestation has taken place in recent decades, tropical forests still cover approximately one-third of Xishuangbanna; this prefecture is the only area in China where stands of virgin tropical forest can be found. Flowering plants and ferns number roughly 4,500 species, representing one-sixth of the described taxa for all of China.

Much of southwestern China, including Yunnan Province, is inhabited by ethnic groups of non-Han Chinese status. In Yunnan Province, twenty-three different ethnic groups are officially recognized as "national minorities." After the founding of the People's Republic of China in 1949, the central government set forth the "Policy of National Unity and Equality" to support local economic development and to help maintain and protect minority groups, their languages, and cultural heritages. Under the government's policy of regional autonomy, eight minority national autonomous prefectures have been established since 1950. One of these prefectures is the Xishuangbanna Dai Nationality Autonomous Prefecture.

Ethnically speaking, Xishuangbanna is quite diverse. Of a total population numbering nearly 700,000, 208,000 are Dai and 84,000 are Aini (or Hani). Other ethnic groups—such as the Bulong, Lahu, Yao, Wa, and Jinuo—have populations of not more than 30,000 each, and the Miao and Kucong groups, less than 1,000. More recently, another ethnic group, the Kemo, has been recognized. They number about 800 people and are a Mon-Khmer-speaking group, spread along the border between Yunnan and Laos.

These groups are native to the area and each has its own history and cultural tradition. But the Han Chinese and the Hui[3] nationalities are immigrants from other parts of China, though they have been settled in Xishuangbanna for hundreds of years. The Han Chinese and the Hui population is about 220,000 and 10,000, respectively; many of them are rubber planters and town residents. Thus, the Dai and the Han Chinese constitute two-thirds of the population, while the remaining third is comprised of twelve other minority groups, the smallest of which is the Kucong with about 500 people (Zheng 1981).

As the largest native group, the Dai people have a long history of living in Xishuangbanna (Wang 1979). According to the ancient Chinese text in the *Hou Han Shu* ("Book of the Later Han Dynasty"), modern Dai ancestors lived as early as 200 B.C. in a large area of what is today southwestern Yunnan and have long had a close association with the Han Chinese. The language spoken by the Dai is of the Sino-Tibetan language family—the Zhuang[4]-Dai language branch.

The Dai have their own written language, culture, agricultural system, and traditions of plant use. They live in lowland villages, located along streams and rivers, in houses raised above the ground on poles. The distance between Dai villages is roughly 3 to 5 kilometers and each village normally has 30 to 80 households.

Due to the inaccessibility of the area and its rich natural resources, and existence of a feudal political system, the Dai people developed a relatively self-sufficient way of life and economic system. They had a settled agricultural system, based on the seasonal cultivation of paddy rice, along with homegardening that combined fruits, vegetables, and herbs. In addition, the Dai possessed a traditional pattern of fuelwood cultivation involving growing plantations of *gemaixili* (*Cassia siamea* Lam.).

At the same time, the conception of "Holy Hills," one of the Dai people's cultural beliefs derived from an earlier and formerly more dominant polytheistic religious tradition, helped to preserve certain areas in pristine forest vegetation (Pei 1985). As a result, natural forests were well maintained in such areas, and the wild products from these forests provided a sustained supplement to their overall agricultural system. For example, more than 100 species of wild vegetables and edible fruits were collected from the wild plant population (Pei 1982), 73 different species of timber were gathered from the natural forests (Yu, Xu, and Huang 1982).

Although mountain minorities—such as the Hani, Bulong, Yao, Jinuo, Lahu, and others—also have their own histories, traditions, and spoken languages, few have their own system of writing. These minorities usually cultivate upland rice by slash-and-burn techniques. They made their living in this manner without permanent settlements for thousands of years. While their pattern of cultivation does not seem, to this writer at least, as sustainable as that of the Dai, these upland minorities have long combined diligence and wisdom and developed great skills in cultivating tea trees, raising shellac, and collecting and processing bamboo shoots, rattan, edible fungi, natural drugs, and other products of the forest, making use of many different species of flora and fauna.

The mountain people trade, by barter or in cash transactions, the products they have gathered from the forests with river valley groups—such as the Dai, Han Chinese, and Hui—for grain, cloth, salt, needle, silver, and various decorative materials. This exchange of commodities between upland and lowland peoples has been an important factor contributing to a relatively peaceful coexistence and mutually beneficial relationships between the two groups of peoples (see Figure 7.1).

Historically, the mountain tribes of the Xishuangbanna area also grew opium. Before the 1950s, they traded raw opium, which was extracted from poppies, with the Dai people in the valleys and with Han Chinese vendors from outside Xishuangbanna in exchange for grain and other items. However, after 1949, the People's Republic of China launched what was to be a successful campaign to eliminate opium and other kinds of drug abuse throughout the country (Shen 1984). On February 14, 1950, Premier Zhou Enlai issued an order that strictly banned the use and trade of drugs and the cultivation of drug plants, specifying severe penalties for violations. Practical measures followed; civil administrators, public security and health organizations, and the courts cooperated. Opium dens were closed, and land formerly used to grow opium poppy was returned to food production with the government providing seeds and loans to the peasants. These steps brought quick and positive results, and within three years most illegal drug abuse had been effectively eliminated.

In China today, narcotic drugs are only supplied in designated quantities to hospitals and medical and scientific institutes that are licensed by the government to carry out drug-related research.

Plant Products Represented in the
Markets of Xishuangbanna

Changes have occurred in China since the founding of the People's Republic in 1949. One important change in the economic sector has been a state monopoly on the marketing of major farm products such as grain, edible vegetable oils, cotton, and a number of cash crops. A related change has been a government policy of rationing grain and cotton to the residents of China's cities and towns.

The implementation of these policies brought about fundamental changes in the circulation of commodities throughout the country. With this "state economy" occupying an important place in the overall national economy, state-operated commerce eventually spread throughout the country and controls were placed on the purchasing and marketing of major farm products. At the same time, except during the period of the Cultural Revolution (1966–76), the government allowed certain free markets and private businesses to remain as a supplement to the state economy. According to recent figures, free markets now account for slightly more than 10 percent of the national retail sales in the country.

Markets in Xishuangbanna have played an important role in improving rural economic life and in helping to meet the needs of urban consumers during this period. Since 1950, the mutually beneficial relationships between the upland and lowland populations of the area have increased. As the area's population increased, it stimulated farming production and an expanded need for market goods. The result has been a further development of the markets of Xishuangbanna, especially since 1979 when the government passed new economic legislation that accelerated the growth of the nation's private business sector to supplement the state economy.

The markets of Xishuangbanna now can be generally divided into three different types: village markets, community markets, and county-town markets (Table 7.1). Over the past two decades, I visited more than twenty different markets at different seasons to observe and record the various plant products that were available. For example, I visited the market of Menglun community (Type II), where I live, hundreds of times. I visited many other markets— including the county-town markets (Type III) of Jinghong, Menghai, and Mengla; community markets (Type II) of Mengzai, Shaojie, and Ganlanba; and several village markets (Type I)—more than ten times.

During market visits, I collected the following information on available plant products: (1) plant name, (2) use or purpose, (3) source area, (4) whether gathered from the wild or cultivated, (5) preparation, (6) price of unit purchased, (7) ethnic background of collector or vendor, and (8) name of the village where the collector or vendor lives. Specimens of the plants and plant products were collected during field studies (including field trips and trips to the market) and then preserved at the herbarium of the Yunnan Institute of Tropical Botany, Academia Sinica, in Xishuangbanna. The specimens were identified by the institute's research staff. The institute's investigation of plant products found in local markets is still in progress; therefore, this chapter contains only a preliminary report on some of our findings. We plan to continue and expand the research, along with other ethnobotanical studies.

A detailed list of common, local plant products found in the markets of Xishuangbanna is presented in the Appendix to this chapter. The list includes 170 different species with information on the local and botanical names, uses, source areas, and habitats of each species. Of this total, 106 species are food related (vegetables, 45; fruits, 38; nuts, 9; others, 14); 20, fiber and timber items; 14, spices or dyes; 20, herbal medicines and so forth; and the other 10 are plants associated with cultural activities of the native peoples. The products from these species comprise 70 to 80 percent of the total goods found in the area's markets. This high percentage reflects the relative self-sufficiency of the area's rural economy and that the modern world has yet to have a major influence upon local life, though this is beginning to change. The latter factor suggests the appropriateness of further ethnobotanical studies and the need for them.

One aspect of the list of plant species that should be noted, however, is the relative scarcity of herbal medicines and natural drugs represented in Xishuangbanna markets as compared to other parts of Yunnan and other provinces of China. The reason for this is not entirely clear, but may be partly due to the large number of different ethnic groups, many with long histories of residency in Xishuangbanna. Through time, many groups developed their own knowledge of plants and herbs and their medicinal properties. As a result, most of these products are independently gathered by each group.

Conclusion

I have sought to show how plant products and the markets in which they are sold relate to and reflect interethnic interactions in the Xishuangbanna area of Yunnan Province in southwest China. Markets have long played, and will continue to play, an important role in the local economy and the cultural activities of the many ethnic groups that inhabit this area. As centers for the exchange of farm and forest products between upland and lowland peoples, markets provide the different ethnic groups with a greater variety in plant product choices and also allow for increased interaction among the various ethnic groups.

Overall, the patterns of commodity exchange in the Xishuangbanna markets appear mutually beneficial to both upland and lowland peoples, each group obtaining a portion of their plant product needs from the other group. It seems particularly beneficial for the upland peoples who are able to exchange various forest products for grains, cloth, salt, metals, and other items with the Dai and Han Chinese of the valleys.

This pattern also has certain benefits for the natural environment of the area. Because the upland groups obtain a certain portion of their foods, such as grain, through trade with the valley-dwelling Dai, there has been reduced pressure, through a reduction in the slash-and-burn of agricultural fields, on the mountain forests. This, in turn, has provided the irrigated paddy farmers of the valleys with a more stable water supply.

Throughout this chapter, I have emphasized the diversity of production strategies employed by the ethnic groups of Xishuangbanna. This diversity and its overall benefits for the local economy have important implications for development of the local economy.

Acknowledgments

I wish to acknowledge Drs. N. Jamieson and G. Lovelace of the East-West Environment and Policy Institute, Honolulu, and Dr. Karl Hutterer of the University of Michigan Center for South and Southeast Asian Studies, Ann Arbor, who read, commented on, and provided editorial assistance on earlier versions of this chapter.

Notes

1. During much of the period before 1949, Xishuangbanna was included in Puer prefecture.
2. The Bai nationality, one of China's ethnic minorities, now has a population of more than 1 million people, most of whom reside in western Yunnan.
3. The Hui nationality is the major Islamic ethnic group in China and is mainly located in the northwestern part of the country.
4. The Zhuang nationality, the largest national minority in China with a population of some 14 million, is primarily found in the Guangxi Zhuang Autonomous Region and in the Yunnan and Guangdong provinces.

References

Bye, R.A., and E. Linares
   1983 The role of plants found in Mexican markets and their importance in ethnobotanical studies. *Journal of Ethnobiology* 3(7):1–13.

Pei, S.J.
1982 A preliminary study of the ethnobotany of Xishuangbanna. In *Collected Research Papers on Tropical Botany*. Kunming, China: Yunnan Publishing House. Pp. 16–30. (in Chinese)
1985 Some effects of the Dai people's cultural beliefs and practices upon the plant environment of Xishuangbanna, Yunnan Province, southwest China. In *Cultural Values and Human Ecology in Southeast Asia*, edited by Karl L. Hutterer, A. Terry Rambo, and George Lovelace. Ann Arbor: University of Michigan Center for South and Southeast Asian Studies, Paper No. 27. Pp. 321–339.

Shen, J.C.
1984 How China ended drug abuse. *China Reconstructs* (North American edition) 33(3):27–29.

Wang, J.
1979 The Dai nationality. In *Collected Research Papers of the Yunnan Institute of History*, Vol. 2. Kunming, China: Yunnan Institute of History. (in Chinese)

Yu, P.H., Z.F. Xu, and Y.L. Huang
1982 Research on the ethnical timber utilizations in the district of Xishuangbanna. In *Collected Research Papers on Tropical Botany*. Kunming, China: Yunnan Publishing House. Pp. 108–115. (in Chinese)

Zheng, L.
1981 *Travels through Xishuangbanna: China's Subtropical Home of Many Nationalities*. Beijing: Foreign Languages Press.

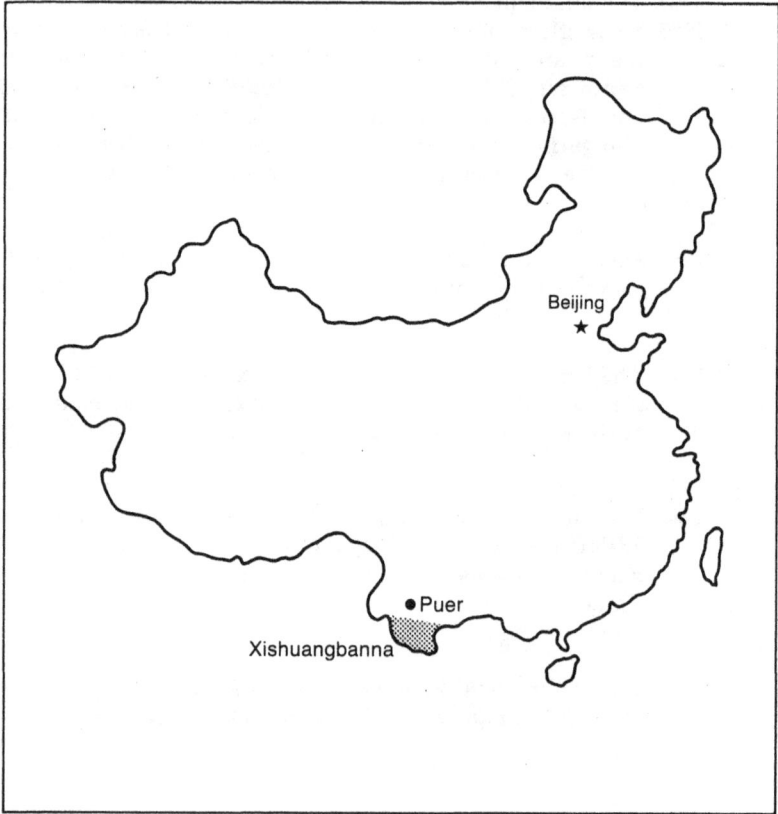

**Map 7.1
Location of Xishuangbanna and Puer in
Yunnan Province, southwestern China**

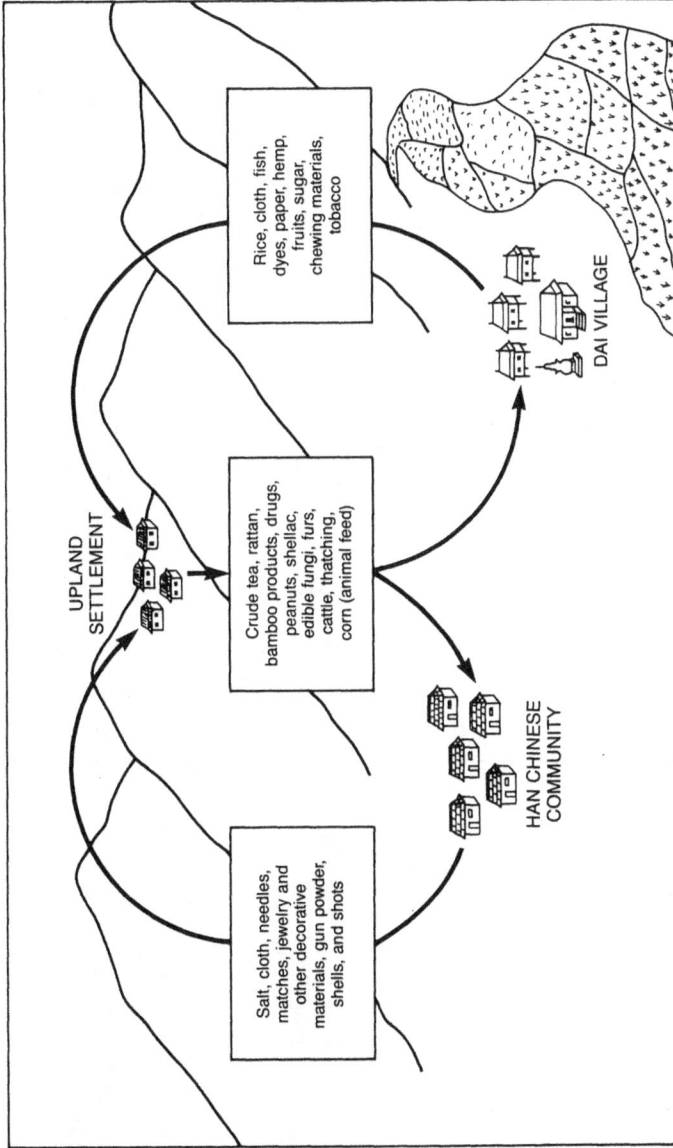

Figure 7.1
Schematic representation of commodity exchange between upland and lowland groups in Xishuangbanna

Table 7.1
Types of markets in Xishuangbanna

	Type I (village market)	Type II (community market)	Type III (county-town market)
Location	large villages	commune centers or small towns	county towns
Number of villages included in marketing area	3–5	20–30	50–80
Number of sellers per day	20–30	200–300	300–500
Number of buyers per day	60–80	500–1,000	1,000–3,000
Market day	Sunday	Sunday and Wednesday	daily
Plant products sold (items/species)	30–50/20–30	150–200/ 100–150	200–300/ 150–200
Source area (km^2)	5–10	10–20	40–50

Appendix: List of common plant products represented in Xishuangbanna markets

Local name (language)*	Botanical name	Uses	Source area	Habitat
Shuimian (H)	Spinogyra cladophora	Food	Valleys	River
Gegu (D)	Cibotium borometza	Herb	Mountain forests	Wild
Paguteng (D)	Pteridium aquilium	Vegetable	Valleys	Wild
Pakezuo (D)	Ceratopteris siliquelosa	Vegetable	Valleys	Wild
Deci (J)	Cycas siamensis	Herb	Mountains	Wild
Songsu (H)	Pinus Khasiana	Timber	Mountains	Wild
Woniaixi (A)	Gnetum montanum var. megalocarpa	Nut	Mountains	Wild
Mahan (D)	Michelia hedyosperma	Spice	Mountains	Wild
Zhangbatum (D)	Paramichelia baillonii	Timber	Mountains	Wild
Magan (D)	Anona reticulata	Fruit	Valleys	Cultivated
Magantulu (D)	Anona squamosa	Fruit	Valleys	Cultivated
Genashabanea (D)	Cananga odorata	Flower	Valleys	Cultivated
Maizhong (D)	Cinnamomom porrectum	Camphor	Mountains	Cultivated
Shahaiteng (D)	Litsea cubeba	Spice	Mountains	Wild

*Language Key: (A) = Aini; (B) = Bulang; (D) = Dai; (H) = Chinese, in local dialectic pronunciations; (J) = Jinuo; (L) = Lahu; (W) = Wa.

Local name (language)*	Botanical name	Uses	Source area	Habitat
Huzhengcao (H)	*Anemone filisecta*	Herb	Mountains	Wild
Jinglang (D)	*Nigella glandulifera*	Herb	Valleys	Cultivated
Nuozhang (D)	*Nelumbo nucifera*	Vegetable	Valleys	Cultivated
Palang (D)	*Parabaena sagitata*	Vegetable	Valleys	Wild
Pake (D)	*Piper sarmentosum*	Vegetable	Valleys	Wild
Jingbai (D)	*Piper betel*	Chewing	Valleys	Cultivated
Yuxincai (H)	*Hauttuynia cordata*	Vegetable	Valleys	Cultivated
Pagong (D)	*Crataeva roxburghii*	Vegetable	Mountains/valleys	Wild
Piela (H)	*Brassica caulorapa*	Vegetable	Mountains/valleys	Cultivated
Kucai (H)	*B. integrifolia*	Vegetable	Mountains/valleys	Cultivated
Bouxiong (H)	*B. oleracea*	Vegetable	Mountains/valleys	Cultivated
Baicai (H)	*B. pekinensis*	Vegetable	Mountains/valleys	Cultivated
Loubu (H)	*Raphanus sativus*	Vegetable	Mountains/valleys	Cultivated
Bocai (H)	*Spinacia oleracea*	Vegetable	Mountains/valleys	Cultivated
Cihancai (H)	*Amaranthus spinosus*	Vegetable	Valleys	Wild
Honghancai (H)	*A. tricolor*	Vegetable	Valleys	Cultivated

Local name (language)*	Botanical name	Uses	Source area	Habitat
Gomasung (D)	Averrhoa caramola	Fruit	Valleys	Cultivated
Meifei (D)	Duabunga grandiflora	Timber	Valleys	Wild
Dukeshan (D)	Dillenia indica	Fruit	Valleys	Wild
Gemaxie (D)	Bixa orellana	Dye	Valleys	Cultivated
Majing (D)	Flacourtia ramontchii	Fruit	Valleys	Cultivated
Mabamenhong (D)	Benincasa hispida	Vegetable	Mountains	Cultivated
Diangua (D)	Cucumis sativa	Vegetable	Mountains	Cultivated
Xionggua (H)	C. pepo	Fruit	Valleys	Cultivated
Herciga (D)	Gymnopetalum cochinchinense	Vegetable (young fruit)	Valleys	Wild
Miangua (H)	Cucurbita moschata	Food	Mountains/valleys	Cultivated
Majing (D)	Hodgsonia macrocapa	Nut	Valleys	Wild
Hulu (H)	Lagenaria siceraria	Vegetable	Valleys	Cultivated
Sigua (H)	Luffa cylindrica	Vegetable	Valleys	Cultivated
Yangsigua (H)	Sechium edulis	Vegetable	Mountains	Cultivated
Mahoula (D)	Trichosanthus villosa	Food	Valleys	Cultivated
Guishebao (D)	Carica papaya	Fruit	Valleys	Cultivated

Local name (language)*	Botanical name	Uses	Source area	Habitat
Yela (D)	*Camellia sinensis var. assamica*	Beverage	Mountains	Cultivated
Maitoulou (D)	*Schima Wallichii*	Timber	Mountains	Wild
Maguehsiangla (D)	*Psidium guajava*	Fruit	Valleys	Cultivated
Gemanda (D)	*Gorcinia Xanthochymus*	Fruit	Valleys	Wild
Mabong (D)	*G. cowa*	Fruit	Mountains	Wild
Geheifei (D)	*Eriolaena malvacea*	Fiber	Mountains	Wild
Maimefei (D)	*Pterospermum diversifolia*	Fiber	Mountains	Wild
Gebo (D)	*Sterculia pexa var. yunnanensis*	Fiber	Valleys	Cultivated
Gebo (D)	*Sterculia villosa*	Fiber	Valleys	Cultivated
Geluezhong (D)	*bombax malabarica*	Fiber	Valleys	Wild
Mianhua (H)	*Gossypium herbaceum*	Fiber	Mountains	Cultivated
Geboman (D)	*Hibiscus macrophylla*	Fiber	Valleys	Cultivated
Maiyao (D)	*Aleurites moluccana*	Lamp oil	Valleys	Cultivated
Mafai (D)	*Baccaurea ramiflora*	Fruit	Valleys	Wild
Gokai (D)	*Homonoia riparia*	Vegetable	Valleys	Wild

Local name (language)*	Botanical name	Uses	Source area	Habitat
Maihongham (D)	Jatropha curcas	Lamp oil	Valleys	Cultivated
Mahangban (D)	Phyllanthus emblic	Fruit	Mountains	Wild
Doyigo (H)	Docynia delavayi	Fruit	Mountains	Wild
Mei (H)	Prumus mume	Fruit	Mountains	Cultivated
Maotao (H)	P. persica	Fruit	Mountains	Cultivated
Li (H)	Pyrus pyrifolia	Fruit	Mountains	Cultivated
Maixiu (D)	Bauhinia varigata	Vegetable	Mountains	Wild
Gefang (D)	Caesalpinia sappan	Herb dye	Valleys	Cultivated
Gemeixili (D)	Cassia siamea	Firewood	Valleys	Cultivated
Mahan (D)	Tamarindus indica	Fruit	Valleys	Cultivated
Xixie (D)	Acacia catechu	Chewing	Valleys	Cultivated
Pana (D)	A. intsia	Vegetable	Valleys	Wild
Gelougai (D)	Sesbania grandiflora	Vegetable	Valleys	Cultivated
Mudou (H)	Cajanus fulvus	Food	Mountains	Cultivated
Tugua (H)	Pachyrhizus erosus	Fruit	Mountains	Cultivated
Sijidou (H)	Phaseolus vulgaris	Vegetable	Valleys	Cultivated

Local name (language)*	Botanical name	Uses	Source area	Habitat
Wandou (H)	Pisum sativum	Vegetable	Valleys	Cultivated
Silungdou (H)	Psophocarpus tetragonolobus	Vegetable	Valleys	Cultivated
Ge-gen (H)	Pueraria edulis	Herb food	Mountains	Wild
Jiangdou (H)	Vigna sinensis	Vegetable	Valleys	Cultivated
Magelue (D)	Castanopsis hystrix	Nut	Mountains	Wild
Magerong (D)	C. indica	Nut	Mountains	Wild
Magebian (D)	C. mekongensis	Nut	Mountains	Wild
Maihapa (D)	Celtis wightii	Nut	Mountains	Wild
Gemaleng (D)	Artocorpus heterophylla	Fruit	Valleys	Cultivated
Gowa (D)	Ficus auriculata	Vegetable	Valleys	Wild
Gemalelang (D)	F. callosa	Vegetable	Valleys	Wild
Gemanlede (D)	F. glomerata	Vegetable	Valleys	Wild
Buma (A)	Ficus pubigera	Fruit	Mountains	Wild
Padie (D)	Celastrus paniculata	Vegetable	Mountains	Wild
Qiekulu (A)	Pyrularia edulis	Nut	Mountains	Wild
Magu (D)	Sclerpyrum wallichianum	Nut	Mountains	Wild

Local name (language)*	Botanical name	Uses	Source area	Habitat
Madian (D)	*Zizyphus mauritiana*	Fruit	Valleys	Cultivated
Mamuo (D)	*Elaeagnus conferta*	Fruit	Valleys.	Cultivated/wild
Mabu (D)	*Citrus grandis*	Fruit	Valleys	Cultivated
Xiongyen (H)	*Citrus medica*	Fruit	Mountains	Cultivated
Magin (D)	*Citrus reticulata*	Fruit	Mountains/valleys	Cultivated
Magin (D)	*Citrus sinensis*	Fruit	Mountains/valleys	Cultivated
Zana (A)	*Evodia trichotoma*	Spice	Mountains	Cultivated/wild
Mageng (D)	*Canarium album*	Fruit	Mountains	Wild
Maiyong (D)	*Toona ciliata*	Timber	Mountains	Wild
Abiu (J)	*Nephelium chryseum*	Fruit	Mountains	Wild
Maiga (D)	*Pometia tomentosa*	Timber	Mountains	Wild
Mamer (D)	*Choerospondias axillaris*	Fruit	Mountains	Wild
Magu (D)	*Dracontomelon macrocarpum*	Nut	Valleys	Wild
Mamou (D)	*Mangifera indica*	Fruit	Mountains/valleys	Cultivated
Galiluo (D)	*Spondias pinnata*	Fruit	Mountains/valleys	Wild
Suilahe (L)	*Aesculus panduana*	Herb	Mountains	Wild

Local name (language)*	Botanical name	Uses	Source area	Habitat
Papimangmang (D)	*Eryngium foetidum*	Spice	Valleys	Wild
Maibushang (D)	*Pouteria grandifolia*	Fruit	Mountains	Wild
Maban (D)	*Eberhardtia yunnanensis*	Fruit	Mountains	Wild
Maban (D)	*Xantolis stenospetala*	Fruit	Mountains	Wild
Pale (D)	*Ardisia solanacea*	Vegetable	Valleys	Wild
Saigang (W)	*Buddleia officinalis*	Dye	Mountains	Wild
Masangduan (D)	*Rauwolfia yunnanensis*	Herb	Valleys	Wild
Gaohahon (D)	*Stelmatocrypton Khasianum*	Herb	Valleys	Wild
Wosun (H)	*Lactuca scariola* var. *sativa*	Vegetable	Mountains/valleys	Cultivated
Gdoumen (D)	*Cordia dichotoma*	Tobacco-rolling paper	Valleys	Wild
Mapi (D)	*Capsicum annum*	Vegetable	Mountains/valleys	Cultivated
Mapinei (D)	*Capsicum minimum*	Spice	Valleys	Cultivated
Sufanqie (H)	*Cyphonandra betacea*	Vegetable	Mountains	Cultivated
Fanqie (H)	*Lycopersicum esculentum*	Vegetable	Mountains/valleys	Cultivated

Local name (language)*	Botanical name	Uses	Source area	Habitat
Yingtao Fanqie (H)	L. esculentum var. carasiforme	Vegetable	Valleys	Cultivated
Maheileng (D)	Solanum coagulens	Vegetable	Valleys	Wild
Xilitong (D)	Nicotiana tabacum	Tobacco	Valleys	Cultivated
Kongxincai (H)	Ipomoea aquatica	Vegetable	Valleys	Cultivated
Shanyoa (H)	Ipomoea batatas	Food	Mountains/valleys	Cultivated
Nuobilong (D)	Mayodendron igncum	Vegetable	Valleys	Wild
Heizhima (H)	Sesamum indicum	Edible oil	Valleys	Cultivated
Huangman (D)	Baphicacanthus cusia	Dye	Valleys	Cultivated
Nuohong (D)	Strobilanthus sp.	Spice	Valleys	Cultivated
Maisuo (D)	Gmelina arborea	Edible flower/ valuable timber	Valleys	Wild
Zayayoma (D)	Elsholtzia blanda	Herb	Mountains	Wild
Jilong (B)	Elsholtzia communis	Spice	Mountains	Cultivated
Layouma (D)	Elsholtzia rugulosa	Beverage	Mountains	Wild
Hommoleng (D)	Mentha haplocalyx	Vegetable	Valleys	Wild

Local name (language)*	Botanical name	Uses	Source area	Habitat
Guangge (D)	*Ocinum basilicum var. pilosum*	Spice	Valleys	Cultivated
Suzi (H)	*Perilla frutescens*	Edible oil	Mountains	Cultivated
Maklian (D)	*Annas comosum*	Fruit	Valleys	Cultivated
Gui (D)	*Musa* spp.	Fruit	Valleys	Cultivated
Magu (D)	*Amomum maximum*	Fruit	Valleys	Cultivated/wild
Gemian (D)	*Amomoum villosum var. xanthoides*	Herb	Valleys	Wild
Maliang (D)	*A. aurantiacum*	Herb	Valleys	Wild
Nuodaihon (D)	*Hedychium coronerium*	Flower	Valleys	Cultivated/wild
Khing (D)	*Zingiber officinalis*	Spice	Mountains	Cultivated
Dongging (D)	*Phrynium capitatum*	Wrapper	Valleys	Wild
Moyu (H)	*Amorphophalus mairei*	Food	Mountains	Wild
Peihen (D)	*Colocasia esculentum*	Food	Valleys	Cultivated
Pabo (D)	*Allium fistulosum*	Spice	Valleys	Cultivated
Jiootou (H)	*A. bakeri*	Spice	Mountains	Cultivated
Manbong (D)	*Dioscorea alata*	Food	Valleys	Wild

Local name (language)*	Botanical name	Uses	Source area	Habitat
Gema (D)	Areca catechu	Chewing nut	Valleys	Cultivated
Wai (D)	Calamus spp.	Rattan	Mountains	Wild
Gebao (D)	Cocos nucifera	Fruit	Valleys	Cultivated
Gegue (D)	Livistona saribus	Fruit	Mountains	Wild
Huangcao (H)	Dendrobium spp.	Herb	Mountains	Wild
Maleimiu (D)	Coix lacryma-jobi	Food	Mountains	Cultivated
Maihaolau (D)	Cephalostachyum pergracile	Basket	Valleys	Cultivated
Gexihai (D)	Cymbopogon nardus	Spice	Valleys	Cultivated
Maibo (D)	Dendroncalamus giganteus	Container	Valleys	Cultivated
Maisong (D)	D. strictus	Bamboo shoot	Mountains	Wild
Yaha (D)	Imperata cylindrica var. malor	Shelter material	Mountains	Wild
Kauan (D)	Oryza sativa	Food	Mountains/valleys	Cultivated
Kaunuo (D)	O. sativa var. glutinosa	Food	Valleys	Cultivated
Nocum (D)	Pleioblastus amarus	Bamboo shoot	Mountains/valleys	Wild

Local name (language)*	Botanical name	Uses	Source area	Habitat
Wei (D)	*Saccharum sinensis*	Sugar	Valleys	Cultivated
Saobacao (H)	*Thysanolaena maxima*	Broom	Mountains	Wild
Kaulong (D)	*Zea mays*	Food	Mountains/valleys	Cultivated
Jiaogua (H)	*Zizamia caduciflora*	Vegetable	Valleys	Cultivated

CHAPTER 8

CHANGING ETHNICITY AND SOCIAL RESOURCES IN A THAI-MON VILLAGE, 1971–81

Brian L. Foster

Ethnicity touches nearly every aspect of social and cultural life. Anthropologists, sociologists, psychologists, economists, historians, political scientists, literary figures, politicians, lawyers, and even bankers all share a lively interest in the topic. It is hardly surprising that the major concepts involving the notion of "ethnic" differences are essentially meaningless without careful definition. There is, in fact, scarcely a common core of meaning in the ways various interested parties use the notion of "ethnic" except that it refers to an important (or potentially important) difference between different kinds of people. The importance of the difference almost always has something to do with distribution of resources. This chapter addresses these core ideas in the study of ethnic relations with particular reference to social and economic changes over the past ten years in a village of Mon traders in Thailand.

The Notion of Ethnicity

All of the diverse approaches to ethnicity can be subsumed under a few general orientations that are closely related to each other. The oldest and most general might be called the "cultural orientation," by which people who share a common culture are grouped together. By this view, the world's population is divided into different "peoples" who are often localized and culturally homogeneous. This common culture is thought to govern people's behavior by providing certain values, norms, and stable personal attributes and dispositions characteristic of the "members" of the culture, allowing them to "understand"

143

one another, for example. Ethnic identities by this view are ascribed, acquired through the processes of socialization and enculturation by which individuals learn their culture. As such, ethnic identities explain why people behave in ways characteristic of their culture and why they have certain values.

The "structural orientation," in contrast, focuses on supraindividual *social* units—on societies, groups, ethnic groups, and other "structural" features of society. In general, investigators working from this orientation are concerned with the relationships among the units. Interest often focuses on distribution of resources among them; on competition, conflict, or cooperation among them; on subordination and superordination relationships; or on absorption of one unit by another. This kind of research is often identified with concepts such as "pluralism," "nation building," "assimilation," and "group conflict." Acculturation studies might be seen as a blend of cultural and structural orientations; they are usually based on the assumption that a culture is a society (or at least a subsociety), which gives meaning to the notion of cultures coming into contact with one another.

In the third orientation, which I will call the "social orientation," we find a comprehensive and theoretically interesting unification of cultural and structural ethnicity, along with various social psychological ideas. Scholars with this orientation focus on the social uses of culture and social structure. I think it is here that we find ethnicity related to resource allocation in the most interesting way. The relationship is simply that the uses of culture and social structure often (perhaps always) concern the distribution of resources, which may range from political support to wealth and on to information or even privilege. A good deal of social life consists of individuals and groups of individuals allocating resources. The crux of such processes is the way claims are validated and denied, and the principles by which allocations are made. It is in this regard that ethnic relations become implicated as, for instance, when the right to resources is claimed by virtue of having a certain ethnic identity.

To see in more detail how this works, it will help to look at the different orientations from a slightly different perspective. Cultural ethnicity is cultural very much in the way of old-fashioned cultural anthropology. Units of analysis are cultures, systems of symbols, meanings, beliefs, and other forms of learned behavior. Structural ethnicity, in contrast, is social-structural; units are societies, groups, categories of individuals, social relations, roles, and so on. From the social ethnicity perspective, the integration of cultural and structural ethnicity is based on the fact that all people recognize correspondences between the distribution of cultural elements and the boundaries of social units, and that they interpret such correspondences. The term "ethnic" concerns the interpretation and use of those correspondences. The salience of the correspondence between social and cultural boundaries depends on many conditions, but in cases where it becomes salient, ethnic relations assume importance in structuring social and/or cultural life. There are at least two ways of conceptualizing the saliency of ethnicity: on the one hand, it can be seen as the "degree of regularity with which any specific role is ascribed on the basis of ethnic identities" and, on the other hand, it is "the degree to which ethnicity structures interaction for all the different roles in a society—the numbers of roles ascribed on the basis of ethnicity" (Wu and Foster 1982:288).

In general, ethnicity is salient when it governs or influences circumstances where something important is at stake—when ethnic differences regularly enter into role definition and thus become relevant to distribution of resources. Consider an abstract example. A resource is controlled by people of a cultural type, who also constitute a social unit. A person may claim access to a resource by claiming affiliation with the social unit that controls it; that claim to affiliation may be supported by showing cultural traits appropriate to the unit's "members." An actual case in point is seen in American Indian tribes' land claims or claims in certain social legislation. For an individual to claim such resources, he or she must claim membership in the tribe (i.e., an appropriate ethnic identity). The membership claim must be supported by certain culturally defined traits, most commonly genealogical, but sometimes others as well (e.g., see Hicks and Kertzer 1972). In recent years, it has often been argued that the modern state societies have actually increased the saliency of ethnicity by increasing the range of resources for which people may compete, and for which ethnic mobilization may be effective (Cohen 1978:397).

What makes such cases characteristically "ethnic" is the cultural interpretation of the correspondences between cultural and social units' boundaries. This interpretation, which we might say defines an ethnic boundary, becomes incorporated into symbolic systems that lend it strong affective dimensions. Rights and obligations are often assigned to unit membership and are integrated with the symbolic, affectively loaded interpretations of the boundaries. Obligations may include behaving in culturally appropriate ways, and social control mechanisms may assure members' rights. In all of this, the most important point is that the heart of ethnic relations is in the cultural constructions put on the correspondences between cultural and social differences. The "primary" social and cultural differences in and of themselves are of secondary interest. Accordingly, it is not ethnic groups that are the focus of attention, but rather the social use of ethnic categories, of the symbolic and cognitive representations of cultural and social differences, real or alleged. Ethnic boundaries are not cultural and social differences, but cultural representations of the differences.

From this perspective it is possible to quickly outline the basic processes of ethnic relations that are the subject of this chapter. Ethnic categories are associated with widely recognized rights and obligations and thus take on properties of roles. To claim these rights (or avoid the obligations), individuals claim association with the appropriate category (or claim not to be associated).

To make good on the claim, some kind of validation is necessary. The form of the validation is to show that the claimant possesses certain cultural traits that are recognized markers of ethnic identities, the most common being genealogical, linguistic, political, and religious. To the extent that the traits are appropriate, the claim is validated, and the party to whom the claim is made is compelled to accept it (insofar, of course, as any culturally based claim must be accepted). The strength and nature of the necessary evidence depend on many circumstances, particularly the value (or cost) of the resource at stake.

One of the most important aspects of this orientation is that it defines processes that can produce the most important conditions studied under the rubric of ethnic relations: assimilation, pluralism, acculturation, and artful dodges such as "symbolic ethnicity." Equally important, it can accommodate many conditions that appear to be anomalous from other perspectives. The most important are, perhaps, multiple ethnic identities (i.e., a single individual claims different identities in different circumstances, a condition sometimes called "situational ethnicity") and changing identities (e.g., a Chinese changes to become Thai). These are precisely the kinds of conditions and processes that must be addressed to understand the ethnic relations of the Mon traders in Thailand.

Ethnicity and Commerce

This discussion will look at minority trading populations, or middleman minorities as they are often called, in an unusual way. The topic is most commonly approached by investigating ways in which the minority culture fosters successful entrepreneurship. The best known such arguments, of course, follow closely the lines of Max Weber's analysis of the Protestant ethic, building on the proposition that certain cultural beliefs and values promote capital accumulation, profit motive, or some other element of successful entrepreneurship. Other approaches to the problem also generally focus on the content of the minority traders' culture, though perhaps in a different fashion. An example of particular relevance to the Thai case is Edna Bonacich's (1973) analysis of sojourning traders. Consistent with the approach to ethnicity outlined earlier, however, this discussion focuses not on the content of the minority traders' culture, but on the social use of the different cultural identities of the traders and the people with whom they trade.

The key to understanding the Mon traders from this perspective is to understand the nature of commerce, particularly how it fits into traditional societies. Operationally, the question is, what kinds of rights and obligations must traders and their customers observe if their trade is to be economically viable? And what kind of role must the trader assume in order not to disrupt traditional community social organization? In other words, how must traders' roles be defined to make trade viable, and how can such roles fit into traditional communities? And, of course, how do these issues relate to ethnic relations?

The crucial property of trade for our purpose here is that trade in necessities, especially if it requires a large portion of the parties' total resources, is highly stressful. This is so, since the buyer depends on the transactions to acquire something he needs, and the trader depends on profits from the transaction to make his livelihood. Interests are inherently opposed, and one party's gain is the other's loss (see Foster 1974, 1977a, 1977b). This is potentially a dangerous kind of relationship that must be carefully managed if the trader is to pursue his livelihood in a viable manner, and if the harmony of communities in which trade appears is to be preserved.

The problem is all the more difficult in traditional societies, in which local trade for necessities is uncommon, and roles suitable for traders have not been culturally defined. Formal mechanisms for keeping order are unlikely to be able to deal satisfactorily with the peculiar problems of such relationships, since traditional norms are inappropriate. In any event, traditional norms for generosity and fair dealing, especially with one's friends, neighbors, and kinsmen, are sure to be violated as the parties to the trading transactions pursue their opposed interests. It will be difficult in most cases, for instance, for the trader to collect debts in the face of obligations to show generosity and render aid to those in need. Overt bargaining and other typically commercial behavior directed to "getting the best deal" will be negatively sanctioned.

All these conditions are true for villages in Thailand. Generosity and fair dealing with one's co-villagers are valued and stand in direct contradiction to commerce. Formal social control mechanisms tend to favor the trader's position over the traditional, anticommercial tendencies of Thai culture, but they are not effective. Anticommercial tendencies are strengthened by the characteristically Thai avoidance of face-to-face conflict or unpleasantness of any kind (Phillips 1965, 1967), which make haggling and other trading activity stressful departures from polite social intercourse. In fact, Thai villagers even today view commerce as a selfish, impolite, and downright dishonest kind of behavior, in strong contrast to such other occupations as, say, government employment or farming (Foster 1974, 1977b).

The role of ethnicity in these circumstances is this: Being of a different ethnic "group" than one's customers is a kind of preadaptation to trade insofar as it establishes social distance that mitigates the stress and conflict inherent in trade. Being an outsider, especially one so distant as to have different ethnic identity, frees the trader from traditional expectations of generosity and fair dealing that are so contrary to the viable pursuit of trade. Moreover, it removes the trader from the effects of many traditional mechanisms of social control, which rely on personal relationships that the trader as an outsider does not have with other villagers. In the Thai case, it frees the trader from the expectations of normal polite social interaction, allowing him to function under the villagers' expectation that he will not behave properly in any kind of relationship (Foster 1977b). In short, the kinds of social relationships characteristic of outsiders (especially of a different ethnic identity) are essentially the same as those required for commerce.

If this line of reasoning is correct, one would expect to find that individual minority traders would be more successful in their trade than those who were not separated by an ethnic difference. They should, therefore, become more numerous in trade than members of the host majority but, at the same time, should be held in low esteem. In 1971 I studied this topic in Thailand among the Mon, a little-known minority whose members had achieved remarkable success in one limited part of the commercial economy—the trade in utility ceramics. The case is particularly interesting in two ways. First, the Mon were culturally similar to the Thai among whom they traded, a situation that made the culture-based theories of entrepreneurship of questionable relevance. Second, most of the trade in Thailand is controlled by Chinese, a group well known as an overseas trading minority.

Very briefly, the Mons' situation was this. The ancestors of the present Mon population migrated to Thailand from southern Burma in the eighteenth and nineteenth centuries, settling in their own villages primarily in central Thailand near Bangkok. Most were rice farming peasants like most Thai villagers around them, and by the early twentieth century, people in these villages had gone far toward assimilation by the Thai. By 1970 they were culturally Thai and regarded themselves as fully Thai in every way. The traders, in contrast, had retained their Mon ethnic identities and many elements of Mon culture, even to the extent that the Mon language was the everyday medium of communication in the village of traders where I worked in 1970-71.

Other aspects of the Mon trading community were even more telling. Most important, in the village of Mon traders, shops and other local commercial establishments were owned by non-Mons—in fact, for the most part by Thais, who seldom engage in trade in their own villages (Foster 1974). But the Thais who had married into the Mon village and taken up the pottery trade, which was done from boats primarily in Thai villages, became Mon—assimilation that clearly went against the prevailing tendency of Mons to become Thai. In other Mon villages (i.e., villages with Mon ancestry but not with the Mon trading occupation and showing little evidence of traditional culture), traders who really made a living from their trade similarly tended to be Thai or Chinese and not Mon, even though the Mon language and other cultural elements had for the most part disappeared, leaving only a kind of "symbolic" Mon ethnicity (Foster 1982). However, those traders who pursued strictly marginal trade, such as selling sweets to school children, showed no such tendency, being approximately equally likely to be Mon or not (Foster 1982:62-64). Moreover, the traders tended to be socially isolated from the other villagers in precisely the ways predicted by the theory (Foster 1980). The theory, then, seems to be supported by the 1971 data.

Design of the Present Analysis

Although all the evidence in my 1971 data seem to support my arguments on the Mon traders, there is a serious (but at that time unavoidable) flaw in the design. This flaw stems from the fact that the 1971 data are cross sectional and do not directly get at the processual nature of ethnic relations postulated in the theory. A truly adequate test of these ideas would require a pretest of a group of people, some of whom were traders and some not. Following the pretest, a significant number in each occupational subgroup must change occupation over a suitable period of time—say, ten years. A post-test, then, would determine whether the changes in occupation co-occurred with those of ethnicity in conformity with the theory. Several predictions, which could serve as the basis of a sound test of the theory, could be made: (1) One would expect to find that Thai who had become traders in the Mon social context would become Mon; Thai who did not enter commerce would not be expected to change. (2) Mon who left trade would become Thai, while those Mon who remained in trade should remain Mon. It would be helpful to control for variables that might affect assimilation—e.g., young people might be the ones most likely to assimilate. Such a design would constitute a much stronger test of the hypotheses than would be possible with only cross-sectional data, but it would make extraordinarily heavy demands on data.

In 1981 I returned to Ban Klang, the Mon traders' village where I worked in 1971, and collected data roughly comparable to that from the earlier fieldwork. Conditions had changed considerably in the village, for the most part in ways that approximate the conditions needed to test the "theory" of Mon traders. Most important, although many villagers remained in the pottery trade, many others had changed occupation. Actually this was not surprising, since in 1971 there was evidence of the decline of the pottery trade due to a complicated set of conditions, including competition from truckers. At that time the predominant new occupation that seemed to be emerging was trade in sand, which was dredged from the Chao Phraya River near Ban Klang. The people bought enormous barges, bought sand from the owners of the dredges, and sold it in Bangkok to brokers. It was clear even then that this new occupation had a limited future, since the supply of sand in the river was being rapidly depleted, and since the building boom in Bangkok was diminishing and with it the demand for sand. By 1981 there were no sand traders in Ban Klang.

The importance of the sand trade was twofold. First, it provided a transition or entry into what was to become the new, emerging village specialty—hauling goods for hire in barges. Many villagers acquired barges in the course of the sand trade, and it was natural to change to hauling for hire when the sand trade ceased between 1971 and 1981. In fact, there were already a handful of barges carrying for hire either full or part time in 1971. Second, the sand trade continued the tradition of commerce in the village, and in 1971 Ban Klang remained predominantly a trading village even though the pottery trade was in decline. In this regard, then, it is extremely interesting that Mon ethnicity in Ban Klang began to decline dramatically from 1971 to 1981. The Mon language was spoken noticeably less around the village. Perhaps the most important change was that the young children, who in 1971 had spoken Mon among themselves, used Thai almost exclusively in 1981.

It appears, then, that together the 1971 and 1981 data substantially meet the conditions for a proper test of the theory of the Mon traders' ethnicity. There are several limitations, however, which must be addressed before looking at the design more closely. First, the fieldwork in Ban Klang met with difficulties both in 1971 and, especially, in 1981. The villagers were not as open and as trusting as in other villages, partly, I suspect, because their occupation made them vulnerable to police corruption and other problems. They were never sure just why I was there. But more than that, special problems arose on both occasions. In both years, interviewing was done about the time of the Songkran festival, the one occasion during which most of the traders returned to the village at the same time. In 1971, one of the interviewers proved to be unreliable and simply invented much of the "data." His work had to be redone, but by the time I got to that, many of the boatmen had left to do their trading, and I had no further access to them. In 1981 the long-time abbot at the temple died shortly after we arrived. It was impossible to do any interviewing during the long and elaborate funeral ceremonies; when they ended, most of the boatmen left before we could contact them.

As a result, a great deal of data is missing for this village. In 1971 Ban Klang contained 80 households, with about 100 nuclear families, and in 1981 about 100 and 150, respectively. For this chapter I have used the 1981 data only on the forty-five heads of families who appear in both data sets and for whom we have no missing data on ethnicity or occupation. This is a heavily biased sample, since a disproportionate number of young people who run barges and haul goods for hire had left to work, while housewives, retired people, and villagers following "usual" occupations were overrepresented among those who remained in the village. Nevertheless, the sample of forty-five includes thirteen people who sell pottery by boat, sixteen who are in nontrading occupations, and twelve who live at home and do not work at any "outside" occupation. The remaining four are "local" traders who carry out their trade within the village. Although far from ideal, this diversity is sufficient for substantial controls on major sources of bias in the sample.

The second limitation is that samples are small, and occupational changes are distributed nonrandomly over the population. For example, the number of people becoming traders during the 10-year period is too small to get statistically meaningful results; virtually no "real Thais" enter trade in ceramics. This places strict limits on our ability to control effectively for variables that may affect assimilation. It is not possible, then, simply to regard the "treatment" in this "experiment" as changing occupation; rather, the situation needs to be redefined slightly to take best advantage of the particular data at hand. The situation is that there is in Thailand a strong, pervasive force for assimilation of Mons (and others, it might be added, including the Chinese). The theory predicts that the Ban Klang Mons will assimilate in exactly the same way as others with an important exception. Being engaged in commerce is a countervailing force to assimilation, implying that Mons in Ban Klang should be assimilating except for those who remain in the pottery trade or other trade with non-Mons. In the modified design, then, we will expect to find those in trade remaining Mon, while those not in trade tending to become Thai.

We can also address the question of causal direction in a limited way. The question is, does the change in ethnicity precede the change in occupation or follow it? If ethnicity changed first, we would expect to see Mons who changed to Thai, but who were still in trade. By similar reasoning, if the change in occupation precedes the change in ethnicity, we would expect to find former traders who are still Mon. We can address this issue even in this limited way only because we have data from two points in time.

Several other complexities resist any systematic solution given the nature of the data, but they must be mentioned before turning to the data. Few people in Ban Klang were not traders at some time in the past, which raises the complication that we cannot simply ask whether a person was once a trader, but must ask how long it has been since he was a trader. Small samples make it impossible, however, to control for any length of time in or out of trade in any systematic way. Actually, given the nature of the process linking ethnicity and commerce, one would expect a time lag between change of occupation and change in ethnic identity. The process is a kind of selection in which individuals who do not have the appropriate ethnic identity are selected out of commerce, while those who do tend to survive longer, all other things being equal. What is going on here, strictly speaking, is that those who have become assimilated have trouble in trade and tend to fall out of it, while those who do not assimilate tend to do better and therefore to become preponderant in the population of traders. Those who changed occupation long ago, then, would be expected to have changed ethnic identity, while those who had recently changed may or may not. For that matter, the same would apply to those who had changed before 1971, further complicating the situation.

Two final elements of the situation require some comments. First, some traders trade within the village. They require special consideration, since trade with "one's own people" is different from selling pots to Thais, and the implications of the theory for these people must be carefully examined. Local traders whose commercial activities take place in a Mon population would not assume a Mon identity (Foster 1974). The condition that promotes successful trade is that the trader's identity be different from the customer's, and if the customers are generally Mon, the trader must assume another identity. One might argue that the Mon identity of the people of Ban Klang has become sufficiently problematic that this condition would not be met. I would argue, though, to the contrary, that the appropriate condition does still obtain in 1981, since most of the people who are in the village most of the time are the old people who retain their Mon identities. The younger people who have taken up the barge transportation occupation are seldom in the village and, in any case, are considerably less visible and socially important than their elders.

Second, the people with no occupation are not directly accommodated by the theory, and some position needs to be taken about them. In the absence of a good reason to believe otherwise, I assume that no change occurs for those with no occupation. Older, retired people are a special case, since they retain their identification with the occupation they pursued before retirement. But in addition, the old people are those for whom Mon culture is most meaningful, for whom the Mon language may be more natural and fluent than Thai, and for whom being Mon is simply easiest and most natural. Their lives center in the village, where the forces of Thai assimilation are felt but slightly. For them, insulated from the general Thai society, ethnic inertia leads to retaining their Mon-ness.

In summary, then, we have a situation in which a broad tendency for assimilation is seen by the fact that most Mons in Thailand have become Thai. Remaining Mon is problematic and requires explanation. The theory's main prediction is that Mons who carry out trade among non-Mons will retain their Mon-ness, while those who change from trade to other occupations will change their ethnic identities to Thai. We have few or no cases of non-Mons entering trade or of Mons who were never in trade. Auxiliary propositions are that local traders will become Thai or assume another non-Mon identity, and that old, retired people, most of whom were traders when active, will remain Mon.

Results

The design of the analysis rests on the proposition that a considerable amount of occupational change occurred between 1971 and 1981 and that a considerable number of villagers changed their ethnic identities during that same period. Since much of the data were missing and extremely biased, no tabular presentation of the occupational changes is shown. Occupational data were available for sixty-seven persons. Of these sixty-seven, fifty-three said they were in a different occupation in 1981 than in 1971. Aggregating over the classes "pottery trade," "local trade," "nontrade," and "no occupation," the results are still impressive: more than one-third (27/67) changed category from the one time to the next. Clearly, the requisite condition of occupational change is met.

Similarly, ethnic identities have changed at an impressive rate, as shown in Table 8.1: Approximately one-third of the people changed their identities between 1971 and 1981. Moreover, although assimilation clearly is occurring, in 25 percent of the changes the direction of the change was from Thai to Mon or mixed-Mon rather than from Mon to Thai (or mixed-Mon). These data seem to verify, then, that forces for assimilation are strong, but extremely uneven, in Ban Klang. As suggested earlier, the major surprises are the relatively weak assimilation occurring here (though a significant change toward Thai is noticeable), and the surprising frequency of "reverse assimilation" by which the Thai become Mon. Clearly, an explanation is needed here.

The two classes of occupations that do not directly enter into the hypotheses can be disposed of quickly. Although the four local traders are too few to allow statistically meaningful analysis, some interesting observations can be made. Data on all four are consistent with the theory. One Thai man who remained Thai has no Mon ancestry or other cultural reason to have been Mon except for having been a pottery trader in the past. The precise timing is unclear, but he seems to have been a carpenter, after which he took up pottery trade for a few years, left it because he did not do well economically, and then took up weaving and selling basketry to outside brokers. One is an old (over age 70) Mon lady who left the pottery trade for making and selling basketry. There is little reason to expect either to change. The other two, who changed from Mon-Chinese to Thai, were local merchants in 1981, selling a variety of merchandise to the villagers, as they were in 1971. As expected, both responded to pervasive pressures to assimilate and became Thai. The twelve who are in no occupation present a complex variety of circumstances, and the numbers are too small to allow meaningful analysis. The data contain few surprises, however. Eight of the twelve were Mon in 1971; seven remain Mon and one mixed-Mon, suggesting that ethnicity tends to be stable for them as expected.

The hypothesis that nontraders will tend to be assimilated (i.e., to become Thai) is strongly supported. Table 8.2 shows that 15/16 of those people who are

now active in some occupation other than pottery trade were either Thai or
Mon-Thai in 1981. In 1971, eight of these people said they were Mon, but by
1981 six had changed to Thai and one to Mon-Thai, only one remaining Mon.
Of the eight Thai from 1971, seven remained Thai as expected, and one became
Mon-Thai. All six of the Mon who changed to Thai were at one time or another
in the pottery trade. The one who remained Mon is a man of age 43 who owns a
tug boat, which he hires out to tow boats in the general river traffic. This is
much like a commercial enterprise, and although I have no explicit information
on who his customers are, it is likely that they are primarily Chinese and Thai,
but not Mon. This would make his situation similar to that of the pottery
traders, differing only insofar as he is selling a service rather than ceramics. The
more interesting anomaly is the man who went from Thai to Mon-Thai. He, too,
was in the pottery trade at one time, and in 1971 he was a sand merchant who
bought at the dredges and sold to brokers in Bangkok. His father was Chinese in
1981 but Thai in 1971. Perhaps the most interesting question about this man is
why he was Thai in 1971; I am unable to answer it.

 . Support for the hypothesis is strengthened somewhat by looking at the
subset of six villagers who now haul in barges for hire, since this occupation is
replacing the pottery trade as the predominant source of livelihood in Ban Klang
(see Table 8.3). Four of the bargemen were Mon in 1971; three became Thai by
1981 and one Mon-Thai. One Thai remained Thai, and the other is the man
discussed earlier, who changed from Thai to Mon-Thai. The strongest evidence,
however, comes from those ten individuals who have actually changed occupation, having left the trade in pottery at one time or another. Six of the ten were
Mon in 1971; of these six, four became Thai and one Mon-Thai. One remained
Mon; she is the elderly widow who makes baskets and sells them to middlemen
(presumably Thai and Chinese). None of the six Mons who left the pottery trade
violate the expectations. Four others are somewhat less clearly supportive: (1)
One Thai became Mon-Thai; he is the same anomalous individual discussed
earlier. (2) One person, who was Thai in 1971 and in 1981, is interesting because
he left the pottery trade before the 1971 census to enter the sand trade, which,
according to the theory, should have made him Mon even then. He is anomalous
on all counts. (3) A Mon-Chinese man who became Thai became a local trader,
trading with the Mon villagers, after leaving the pottery trade. (4) I have inadequate data on the fourth.

 Most important, though, are those who were in the pottery trade in 1981
(Table 8.4). Here the data are unambiguous: Of the thirteen traders, eleven were
Mon in 1971 and remained so in 1981; one moved from Mon to Mon-Thai. The
one person who was Thai in 1971 became Mon by 1981. Unfortunately there are
too few people entering the pottery trade to examine the effects of such an
occupational change on ethnic identity. Four who were in the pottery trade in
1981 were not in 1971, but three of them had been in this occupation at some

other time. All three were Mon in 1971. The fourth person was a Thai local trader in 1971 and changed his ethnic identity to Mon as he entered the pottery trade. In summary, then, no one changed from Mon to Thai, clearly in marked contrast to the general Mon population, which is assimilating rapidly and has been for many years. Eleven of the thirteen traders remained Mon, one changed from Mon to Mon-Thai, and one even changed from Thai to Mon. The data strongly support the hypotheses.

Conclusion

It may seem as though we have departed considerably from the relationship between ethnicity and the control of resources. I think not. If we are to avoid merely making superficial, after-the-fact empirical observations about the ways that resource distribution follows ethnic boundaries—and it can surprise no one that resource distribution does often follow ethnic boundaries—we must come to a clear understanding of how ethnic relations function. At best this is a murky area in sociology and anthropology, confused by ideology, romanticism, and just plain lack of appropriate theory. Most theory on ethnicity until the past fifteen years or so was rooted either in a naive and exaggerated kind of primordialist ideology, or it was little more than a classificatory exercise by which individuals were placed in categories with no known general properties beyond having a name. In general, what passes for ideas in theories of pluralism, assimilation, and other "classical" approaches to ethnicity is little more than a restatement of the problem the theories are supposed to be addressing.

The task is to understand the processes of ethnic relations sufficiently well to see how they articulate with processes of resource control. Both are extraordinarily complex dimensions of social life. The newer, "social" orientation to ethnicity, associated with such ideas as "situational ethnicity" and multiple and changing ethnic identities, is an advance insofar as it explains ethnic relations by providing relatively well-defined mechanisms that generate the conditions we observe. The suitability of this approach for these purposes stems from including, at least in principle, processes of rational decision-making and other elements usually associated with resource allocation and control. The three crucial elements in ethnic relations are (1) the characteristic social distance across ethnic boundaries; (2) the symbolic, affective dimensions given to the ethnic categories and the cultural traits that mark them for social use; and (3) decisions about which of one's potential identities to claim in a given situation. The distance is relevant to role definition, which may easily link up with resource allocation; to claim an identity is to claim rights associated with a role. The cultural dimension of ethnicity gives psychological force to the norms that enter into the role definition.

In the Mon case, two kinds of resources are at stake: (1) economic resources, including earning a living through trade, and distribution of goods through trade; and (2) social resources—access to a role appropriate to trading for a living. What makes the analysis work is that the role relations inherent in commerce are understood (as in Foster 1977a), thus allowing us to incorporate the formal properties of the roles in a single formulation with the processes of ethnic relations. This is done by analytically joining the social distance of ethnic differences with the distance that socially insulates trader from customer. The ability to do this allowed more or less precise (and counterintuitive) hypotheses to be generated, which in turn allowed a more or less strong test of the entire formulation.

References

Bonacich, Edna
>   1973 A theory of middleman minorities. *American Sociological Review* 38:583–594.

Cohen, Ronald
>   1978 Ethnicity: Problems and focus in anthropology. *Annual Review of Anthropology* 7:379–403.

Foster, Brian L.
>   1974 Ethnicity and commerce. *American Ethnologist* 1:437–448.
>   1977a Mon commerce and the dynamics of ethnic relations. *Southeast Asian Journal of Social Science* 5:111–122.
>   1977b *Social Organization of Four Mon and Thai Villages*. New Haven, CT: Human Relations Area Files Press.
>   1978 Trade, social conflict, and social interaction: Rethinking some old ideas on exchange. In *Economic Exchange and Social Interaction in Southeast Asia*, edited by Karl Hutterer. Michigan Papers on South and Southeast Asia No. 13. Ann Arbor: University of Michigan Center for South and Southeast Asian Studies. Pp. 3–22.
>   1980 Minority traders in Thai village social networks. *Ethnic Groups* 2:221–240.
>   1982 *Commerce and Ethnic Differences*. Papers in International Studies, Southeast Asia Series No. 59. Athens: Ohio University.

Hicks, George, and David Kertzer
>   1972 Making a middle way: Problems of Monhegan identity. *Southwestern Journal of Anthropology* 28:1–24.

Phillips, Herbert
    1965 *Thai Peasant Personality.* Berkeley: University of California
       Press.
    1967 Social contact vs. social promise in a Siamese village. In *Peasant
       Society*, edited by J. Potter, M. Diaz, and G. Foster. Boston: Little,
       Brown and Company. Pp. 346–367.

Wu, David Y.H., and Brian Foster
    1982 Conclusion. In *Ethnicity and Interpersonal Interaction: A Cross-
       Cultural Study*, edited by David Y.H. Wu. Singapore: Maruzen
       Asia Press. Pp. 279–300.

Table 8.1
Ethnicity in Ban Klang, 1971 and 1981

1971	1981				
	Mon	Thai	Mon-Thai	Mon-Thai-Chinese	Total
Mon	20	6	2	1	29
Thai	2	9	2	0	13
Chinese	0	1	0	0	1
Mon-Chinese	0	2	0	0	2
Total	22	18	4	1	45

Note: In Ban Klang, most of the Mons remain Mon, though a few are changing to Thai and various mixes. None of that is surprising except, perhaps, for the small number who become Thai. Much more surprising is that some Thai become Mon and Mon-Thai. In all, more than one-third of the entire village population changed ethnic identity between 1971 and 1981.

Table 8.2
Nontrading residents of Ban Klang

1971	1981			
	Mon	Thai	Mon-Thai	Total
Mon	1	6	1	8
Thai	0	7	1	8
Total	1	13	2	16

Note: For those villagers who are active in some occupation other than trade, seven-eighths of the Thai remain Thai, while three-quarters of the Mons become Thai. The remaining Mon became Mon-Thai.

Table 8.3
Ethnic identities of those who haul for hire

	1981			
1971	Mon	Thai	Mon-Thai	Total
Mon	0	3	1	4
Thai	0	1	1	2
Total	0	4	2	6

Note: Although the sample is extremely small, it is important to note that the pattern of nontraders holds true strongly for those who entered the emerging village specialty occupation. This provides some evidence that the results would not be changed had we been able to collect data on a larger portion of the population now engaged in hauling for hire.

Table 8.4
Ethnic change for the pottery traders

	1981			
1971	Mon	Thai	Mon-Thai	Total
Mon	11	0	1	12
Thai	1	0	0	1
Total	12	0	1	13

Note: The decisive evidence for the theory takes on its importance in juxtaposition with Table 8.3. The pottery traders maintain their Mon ethnicity virtually without exception in the face of broad pressures for assimilation that seem to be affecting people in other occupations. Although it is not statistically meaningful, it is gratifying to see that one trader who was Thai in 1971 had become Mon by 1981.

CHAPTER 9

ETHNIC CONCENTRATIONS:
THE ILONGOTS IN UPLAND LUZON

Renato Rosaldo

Two views have dominated anthropological studies of ethnicity. An earlier view held that ethnicity as a phenomenon deepens as one moves toward the center. A recent view asserts that ethnicity is an emergent phenomenon appearing at social boundaries. These two conceptions neither necessarily contradict each other nor, more important, do they exhaust the empirical phenomenon to be explained. After considering these two views in turn, I shall propose a third conception, involving concentration in social centers, that better explains the social construction of ethnicity among the Ilongots of northern Luzon, Philippines. This culture-specific formulation conceives ethnicity more in relation to middles than edges and more in connection with social than spatial organization.

My view of Ilongot ethnicity has more general theoretical implications. Most studies of ethnicity have ignored cultural practices and social discourse in favor of ethnonyms and systems of ethnic classification. The problem of ethnic identity has been reduced to the question of labeling (How do people determine that "She is an x"?). Lost in the shuffle are the cultural performances and discourses (talk extending beyond the sentence, not to mention the noun, in length) through which ethnic identity is constructed. Though often conceived as a primordial sentiment (as if given in nature or at any rate ascribed by birth), ethnicity clearly is socially constructed, historically conditioned, and culturally mediated. Not unlike religion and witchcraft, ethnicity stands as one in a family of other human fictions (in the sense of "something made" rather than an "untruth" or a "lie") found compelling by certain of its adherents (true believers), less so by others (skeptics), and yet not at all so by others (atheists).

My project is to ask how ethnicity works when it is compelling and, for the time being, to ignore matters of ethnic skepticism and atheism. Arguably, theories of ethnic boundaries have explored the latter questions by conceiving ethnicity as a tool for interactions more instrumental than spontaneously expressive. In my view ethnicity usually is both instrumental and expressive, and theories that oppose the two perspectives have posed a false dichotomy. Instead I hope to explore the kinds of experiences that ground ethnicity, both reflecting it as a historically produced cultural conception and creating it as a culturally mediated historical event. Human actions at once create a sense of ethnic identity and constitute the collective experiences to which this identity refers. Such identities, like the experiences that make and ground them, change through time as they preserve and revise traditional precedents.

The Ilongots, as foragers who hunt deer and wild pig and as swidden horticulturalists who cultivate dry rice and root crops, orient to a landscape of hills, valleys, and river systems. Their central terms for directionality are *yamu* (uphill) versus *peknar* (downhill) and ´*atat* (upstream) versus *tagud* (downstream). Ilongots thus do not inhabit a landscape marked by extreme microecological variations that favor studies of ethnic boundaries in the manner pioneered by Fredrik Barth. In Southeast Asia such microecological variations have derived from altitude (see Leach 1954 on highland Burma) and depth alongside chains of small islands (see Vivienne Wee, this volume; Frake 1980 on the Sulu Archipelago). The Ilongots inhabit an inland region about 90 miles north of Manila, and their landscape is not highly differentiated in either altitude or depth. In addition, the Ilongots until 1974 had been relatively beyond the control of the nation-state; hence, certain forms of hierarchical differentiation were less salient to them than to others.

Ethnicity Deepens Toward the Center

The first view, that ethnicity deepens toward the center, assumes the phenomenon is evenly distributed in homogeneous space. Empirically, this assumption is complicated in ways that allow for factors such as hills, valleys, and waterways, all defined in relation to particular modes of transportation and communication. The idea is that such factors account for the differences between observed distributions of ethnicity and what otherwise would result from ideal conditions of homogeneous space. The landscape's contours, in other words, are treated as if they were deviations from a norm of undifferentiated flatness. These assumptions underlie the age-area hypothesis and the folk-urban continuum of Robert Redfield's *Folk Culture of Yucatan* (1941).

Why should ethnicity deepen toward the center and grow shallow toward the boundaries? Is there any reason to assume that ethnicity resembles a symmetrically hollowed pool of water? The explanation usually offered is that culture contact produces greater and lesser depths of ethnicity. Differing depths of ethnicity result from contact between cultures that are markedly unequal in political power and economic resources. One of the best historical examples of such inequalities has been the Western Europe phase of imperialist expansion. Indeed this historical phenomenon is precisely what most writers have in mind when they speak, usually without mentioning imperialism, about pre- and post-contact periods.

In situations of dominance and subordination, the dominant culture, whether by design or not, often absorbs or assimilates members of the subordinate group. What the dominant group sees as acculturation, members of the subordinate group experience as deculturation. Among Ilongots in northern Luzon, Philippines, for example, the shift from dry to wet rice cultivation is experienced, among other things, as a loss of ethnic identity ("They took up the plow and were no longer Ilongots"). But for lowlanders, this same shift represents a transition from savagery to civilization. Rather than a loss it is perceived as a gain: the acquisition of superior ways of doing things (often culturally represented as rationality as opposed to superstition).

In this conceptual context, the idea of homogeneous space reappears in the notion that assimilation is a function of distance from metropolitan centers (for a failed attempt to apply this notion, see Guthrie 1970; cf. R. Rosaldo 1972). Proximity to the dominant group produces cultural loss for the subordinate group. Once again, the base of homogeneous space can be empirically modified by geographic contours, technology, and the kind and amount of actual contact. The resulting process, depending on where one stands, can be called assimilation, acculturation, deculturation, or all three together. It can involve a loss, a gain, or both.

Ethnicity here resembles a storehouse filled with wisdom, knowledge, beliefs, practices, and values that have been built up and preserved through the generations. The key terms are cumulative and accumulation. This view usually includes a notion that older traditions are the most authentic. Whether or not it is modified by the term salvage, ethnography conducted within this framework can become a form of disciplined nostalgia as the bereaved fieldworker at once witnesses and attempts in writing to create a memorial that marks the passing of traditional society.

Ethnicity, in this view, revolves around issues of both maintenance *and* resistance. Maintenance implies preserving a condition that obtained before contact with a politically and economically dominant group. At the extreme end of this conceptual continuum, ethnicity resembles an artfully woven fabric handed down whole cloth from time immemorial.[1] When one thread is ripped out, the entire fabric begins to unravel. But resistance most often surfaces in studies of revitalization movements. The less spectacular politics of resistance in everyday life at the local level, however, have been relatively slighted in the literature.

• Whether they concentrate on maintenance or resistance, such studies converge in identifying the dominant group as the main agent of change for the subordinate ethnic group. Ethnicity thus becomes static; something to preserve and defend, rather than the more convincing conception of an array of cultural practices from which each successive generation can select, discard, modify, revise, or recombine with traditional, borrowed, or invented practices.

Ethnicity Emerges at the Boundaries

The second view, that ethnicity emerges at the boundaries, appears most seminally in the work of Fredrik Barth, as an inversion of the first view.[2] Like many revisionist theories, Barth's conception became persuasive because it foregrounded phenomena that the earlier view had ignored. His work introduces a healthy skepticism about equating ethnicity and authentic, central cultural values. Ethnicity, in his view, becomes more a strategic presentation of self than an unwitting reflection of the authentic self. Ethnicity appears as a performance, an agonistic display put on for the members of differing ethnic groups. It is an assertion of difference.

In spatial terms, this model simply affirms that the ethnic action is at the boundaries. Away from the borders, it would seem, ethnicity simply is less in evidence. The more one moves toward the interior, according to this hypothesis, the less people display ethnicity because their neighbors, being the same as they, do not beckon them to assert their difference. In the interiors the "us/them" dichotomy has no relevance; hence, it recedes to the background. Ethnicity, like Claude Levi-Strauss's totemism, has no positive content and instead becomes a system for marking differences. Ethnic identity, a thing that groups put on and take off to signify their difference from other such groups, comes to resemble clothes, masks, emblems, or badges. Clearly, the new conventional wisdom has stood the old on its head.

Ethnicity no longer refers to autonomous whole traditions within groups. Instead it designates dichotomous contrastive relations between groups. Thus one finds that Barth discusses "the cultural contents of ethnic dichotomies" in the following manner:

> When defined as an ascriptive and exclusive group, the nature of continuity of ethnic units is clear: it depends on the maintenance of a boundary. (1969a:14)

> The critical focus of investigation from this point of view becomes the ethnic *boundary* that defines the group, not the cultural stuff that it encloses. (1969a:15)

Barth's revisionist conception thus replaces positive content with a formalist definition of ethnicity. Ethnicity is less a matter of experience and compelling primordial sentiments than of display, what Barth calls "the diacritical features that people look for and exhibit to show identity." When Barth lists such features, he includes "dress, language, house-form, or general style of life" (1969a:14). Ethnicity emerges in relational systems of difference and its emblems become at once arbitrary, external, and material.

Perhaps the most creative use of Barth's perspective in Philippine studies has been a paper by Charles O. Frake (1980) in which he regards ethnic labels as a means of classifying or sorting people into "kinds." People who use them usually regard such labels as ascribed at birth and relatively immutable over their lifetimes. A label of ethnic identity entitles its user to attribute a certain social character to the people so designated.

Frake goes on to show how these labels in the Sulu archipelago form a system of classification designating, among other things, differences in social status. The system itself varies in accord with the social position of the person using the term. People in different relative status positions can label their own ethnicity, as well as that of their neighbors, in markedly different ways. In this manner Frake moves beyond a binary "us/them" view of ethnicity and shows how a group's position within the larger system of differences can alter the classificatory grid itself.

Both views conceive ethnicity, whether it deepens in the center or emerges at the boundaries, as a form of social identity that resembles nationality. These views regard ethnicity, like nationality, as a human fatality ascribed at birth and hence beyond the control of individual volition (see Anderson 1983). Curiously, they usually assume that a person has one, and only one, ethnic identity over the course of a lifetime, as if such affiliations resembled an idealized version of monogamy. From this perspective the equally legitimate possible notions of a succession of identities (as if serial monogamy) and simultaneous multiple identities (as if polygyny) appear deviant and require special explanation as exceptions, rather than simply being acknowledged as other versions within the broad range of ethnic phenomena.

When ethnic groups are assumed to be territorial, like the nation-state, they ideally inhabit a continuous span of territory with well-defined borders. Within such territorial expanses, however, ethnicity could exemplify both views by growing more authentic toward the interiors *and* being defended by agonistic displays at the boundaries. One view does not contradict the other. The notion of ethnic groups as quasi-nations, of course, should be applied with great caution, particularly when speaking of periods before the nineteenth century or groups relatively removed from the control of modern nation-states.

The Map is Not the Landscape

Both views of ethnicity conflate the "map" with the "landscape." They derive from the model of a geopolitical map where continuous territory is divided into mutually exclusive nonoverlapping groups. Each discrete group is represented by a different color: purple, blue, green. Such representations attempt to make landscapes, highly variegated as they are in contours, flora, fauna, and human uses, resemble flat sheets of paper divided by clear lines and parceled into continuous units of homogeneous color. The analytical problems posed here involve both how literally such maps should be read and how much they should be taken as the norm from which all else deviates as exception.

The geopolitical map usually reflects the viewpoint of the modern state. In the Philippines, for example, the American colonial regime no sooner imposed its rule than it began sorting its territory into discrete ethnic units with one name for each group. The Bureau of Non-Christian Tribes was:

> charged with the duty of conducting systematic investigations to ascertain the name of each tribe, the limits of the territory which it occupies, the approximate number of individuals which compose it, their social organization and their languages, beliefs, manners, and customs, with special view to learning the most practical way of bringing about their advancement in civilization and material property. (Taft et al. 1901:38)

The state's problem of dividing a map into discrete governable units hardly coincides with the problems of the tribespeople themselves. Tribespeople could enjoy multiple names depending on who is speaking to whom in what context or as simple alternative forms used in more or less free variation, and their territorial boundaries could be blurred or well defined. The analyst should make explicit the structural position or positions assumed in conceiving ethnicity. Analyses of ethnicity in the Philippines should ask both how there came to be so many cultural minorities and how lowlanders came to enjoy political and economic hegemony in the nation.

Ilongot Ethnicity Gathers in Social Centers

Let me now propose a third view of ethnicity, one that freely draws on Ilongot views of their own identity. Indigenous conceptions of ethnicity, though molded by everyday life concerns more pragmatic than conceptual, should inform theoretical writings on the topic. People's self-conceptions inform their conduct in ways related to, but not simply derived from, the objective constraints on their actions. In other words, how people conceive their ethnicity in large measure constitutes it. Since the view developed in what follows derives from a Philippine hunter-horticultural group, it most probably applies, with locally appropriate modifications, to other Southeast Asian societies that subsist by foraging, swidden cultivation, or both (for a comparable case in Kalimantan, see Lowenhaupt-Tsing 1984).

Consider Ilongot notions of where their ethnic identity emerges with greatest force. A key term here is ´*upug* (gather together, concentrate, pile up). Ilongots use ´*upug* to refer, for example, to social gatherings, residential movements, collecting dried meat in chunks for gatherings, and any concentration of people, animals, or things. In speaking about headhunting raids, men describe how their vision focuses (´*upug*) when they look at their potential victim. The opposite of this term is *siwak* (disperse, scatter, separate). Ilongots use *siwak* to refer to the dispersal of people after a gathering, scattering seed, and movements where once coresidential groups separate.

Central cultural conceptions come in families of related terms (see M. Rosaldo 1980; R. Rosaldo 1980). The notion of gathering together is closely related to *kegkeg* (strength) and *keneg* (thick). The idea is that, in gathering together, groups thereby become thick and strong. In such contexts groups also are regarded as endowed with *liget* (anger, passion, energy, vitality). Collectively people enjoy being in a state of healthy vitality and well-being. This state is often epitomized in the "concentration" and "anger" that celebrations after successful headhunting raids are said to produce. They also occur, in varying degrees, on expeditions for fishing, hunting, gathering; during discussions for bridewealth and other forms of dispute settlement; in periods of residential congregation; and in magical spells intended to focus energy for health, horticultural productivity, success in the hunt, and diffuse human happiness and well-being.

Compared with the idea that ethnicity deepens in the center, Ilongot notions of "gathering together" define centers in a variety of ways that are more social than territorial. The place where concentration occurs is wherever people congregate for culturally valorized, focused activity; it is a movable and not a fixed site. The site could even be the lowlands, as it was, for example, during July and August 1945 when Ilongots, who had fled into the Magat Valley to escape the ravages of war, gathered together and celebrated having taken Japanese heads.

In relation to ecology, Ilongots have a number of distinct sites for their focused activities. Most ethnographic analyses largely limit themselves to hearths, homes, houselots, granaries, and gardens. They should also consider, however, the terrain insistently named in casual conversation, in making plans for foraging, and in song and story alike. Both men and women value knowledge of the landscape for its own sake. Indeed such knowledge enhances the social standing of the knower. The larger landscape comprises a repository of named streams, hills, rocks, and trees that serve both to orient humans as they traverse their environment and to memorialize events significant to the local population. The larger landscape so memorialized is wide-ranging and discontinuous. Parts of it, being better known than others, are more densely named, whether because they have been more intensely foraged or more packed with remembered historical events. Maps reflecting such indigenous conceptions of the landscape could contain dots, here more concentrated (´*upug*) and there more dispersed (*siwak*), or tones, ranging from darker to lighter.

Ilongots say they know no boundaries for their gardens, their hunting, their other foraging activities, or their broader territory. Yet they do have a notion that outside forces can *runek* (penetrate) their collective selves. They refer in this manner to lowlanders, other upland groups, and Japanese entering their region. This notion, conceptually, is like their idea of disease penetrating a healthy body. Such penetration is debilitating because it produces a dispersal of health-giving vitality. The best way to prevent debilitation is to make strength symbolically gather together (´*upug*) through magical spells against illness, and through social congregation in the face of lowland intruders.

Reconceiving Ethnicity

The model of ethnicity derived from Ilongot notions departs from one that works by analogy with the nation-state where a continuous homogeneous territory must be defended, or at any rate defined, against outsiders. The dichotomy between ethnic identity and ethnic opposition, ethnicity as expressive primordial sentiment and ethnicity as instrumental tool of interest-group politics, now appears as a distinction more analytical than empirical. Both conceptions can be copresent and the analyst need not necessarily choose one view as opposed to the other.

Indigenous cultural conceptions, along with their modes of socially constructing identity, theorize the ways in which collective experiences both create and reflect ethnicity. These formative experiences follow loosely patterned forms that embody specific historical experiences. Without such experiences primordial sentiments could not be grounded and interest-group politics could not be sustained over the long run. From this perspective, ethnicity should be theorized in relation to an array of culture-specific conceptions and practices. Like marriage, family, and descent, conceptions of ethnicity emerge less as identities than as family resemblances across a range of cultures. This theoretical perspective revises the earlier two views of ethnicity in at least three distinct ways.

First, ethnicity, for members of such groups, constitutes a fundamental identity as much as a diacritical emblem strategically deployed for differentiating "us" from "them" in in-group versus out-group fashion. This source of identity "thickens" in relation, not to inert received tradition, but to culturally valorized gatherings. These gatherings reenact past actions but in ways that recollect, select, discard, and improvise on how things have been done in the past. Ethnicity thus becomes more than strategic display and less than sheer repetition.

Second, in connection with locally significant environments, ethnicity appears as discontinuous, thicker here and thinner there. The differentiated landscape has such contours because of its use in subsistence activities and historical episodes. The landscape is endowed with its significance through collective experiences of local groups. What makes Ilongots their ethnic selves, in this respect, is less that they share particular knowledge than that they share general ways of registering and memorializing their experiences. Ilongots recognize the validity and import of the knowledge held by other members of their ethnic group even when it differs in particulars from their own local knowledge. Outsiders remain skeptical, less generous in appreciating what it means to know and celebrate experiences mapped onto the local terrain.

Finally, Ilongots recognize that they can be penetrated by outsiders and lose, or at least dilute, their own identities. Yet this penetration is not a matter of crossing boundaries or lines in space. Instead it is more a matter of entering social gatherings, the concentrated potency that comes of collecting and thickening social identity. Ethnicity thus is constructed and reconstructed in ways that reflect and create a sense of group experiences in an ecologically and historically significant landscape.

Acknowledgments

Field research among the Ilongots in 1967–69 and 1974 was financed by a National Science Foundation predoctoral fellowship, by National Science Foundation Research Grants GS–1509 and GS–40788, and by a Mellon award for junior faculty from Stanford University. I am grateful to Mary Louise Pratt and the other participants in the Michigan conference for their comments on this chapter.

Notes

1. In a justly famous figure, Ruth Benedict uses a Digger Indian proverb to compare culture with a clay cup. Her informant told her: "In the beginning, God gave to every people a cup, a cup of clay, and from this cup they drank their life" (1934:21). Referring to the demise of his own culture, the

man went on to say, "They all dipped in the water, but their cups were different. Our cup is broken now. It has passed away" (1934:22). Elsewhere Benedict follows the figure I have cited and speaks of the weave of culture.

2. Another reading of Barth could show how he stresses communication and core values. Thus, rather than inverting the first view he incorporates it in a more encompassing theory. In either case, my position is that two historically separable views of ethnicity should be combined, and indeed revised in the light of what I call a third, yet more encompassing, view.

References

Anderson, Benedict
> 1983 *Imagined Communities: Reflections on the Origin and Spread of Nationalism.* London: The Thetford Press.

Barth, Fredrik
> 1969a Introduction. In *Ethnic Groups and Boundaries*, edited by Fredrik Barth. Boston: Little, Brown and Company. Pp. 9–38.
> 1969b Pathan identity and its maintenance. In *Ethnic Groups and Boundaries*, edited by Fredrik Barth. Boston: Little, Brown and Company. Pp. 117–134.

Benedict, Ruth
> 1934 *Patterns of Culture.* Boston: Houghton Mifflin Company.

Frake, Charles O.
> 1980 The genesis of kinds of people in the Sulu archipelago. In *Language and Cultural Description: Essays by Charles O. Frake.* Stanford: Stanford University Press. Pp. 311–332.

Guthrie, George M.
> 1970 *The Psychology of Modernization in the Philippines.* Quezon City: Ateneo de Manila University Press.

Leach, Edmund
> 1954 *Political Systems of Highland Burma.* Cambridge: Harvard University Press.

Lowenhaupt-Tsing, Anna
> 1984 Politics and culture in the Meratus Mountains. Ph.D. dissertation, Stanford University.

Redfield, Robert
  1941 *Folk Culture of Yucatan.* University of Chicago Publications in Anthro, Social Anthro Series. Chicago: University of Chicago Press.

Rosaldo, Michelle
  1980 *Knowledge and Passion: Ilongot Notions of Self and Social Life.* Cambridge: Cambridge University Press.

Rosaldo, Renato
  1972 Review of *Psychology of Modernization in the Philippines,* by George M. Guthrie. *American Anthropologist* 74:1439–1442.
  1980 *Ilongot Headhunting, 1883–1974: A Study in Society and History.* Stanford: Stanford University Press.

Taft, W. Henry, D. Worcester, L. Wright, H. Ide, and B. Moses
  1901 *Reports of the Taft Philippine Commission.* Washington, D.C.: Bureau of Printing.

CHAPTER 10

HIERARCHY AND THE SOCIAL ORDER: MANDAYA ETHNIC RELATIONS IN SOUTHEAST MINDANAO, PHILIPPINES

Aram A. Yengoyan

The Problem of Ethnic Group Classification and Ethnicity in Southeast Mindanao

Throughout the ethnographic accounts of the upland peoples of southeast Mindanao, the designation of societies, cultures, and languages has always been characterized by an endless number of particular terms and labels, many of which have almost no reality either in terms of cultural groupings or language families. Since the early work of Blumentritt, most ethnographers have tended to divide and subdivide groups into finer categories whose existence is highly suspect. Over time, the literature on these groups has stressed the idea that these were coherent cultures in a particular locality and that each group could be established by some set of diagnostic features that demarcates one society from another. Although the elaboration and overelaboration of this process is now widespread throughout the upland areas of Southeast Asia and has existed as well in the precolonial period, the colonial impact not only enhanced this process for administrative purposes but also tried to establish a buffer zone between the lowland societies and the uplanders. Since most colonial regimes assumed that the lowlanders had been "contaminated" historically by larger traditions such as Islamism, Buddhism, Hinduism, and the Catholic Church, the uplanders had to be saved from what had transpired on the coastal and lowland areas, which in turn meant that some sort of "pristine" situation should be maintained through direct administrative control.

Southeast Mindanao provides no exception to what happened in other parts of the Philippines under the American administration. In virtually all the early maps of the people and cultures of the regions, the coastal groups are homogeneously labeled Bisayan speakers, while the uplanders are fragmented into subcategories that stress the differences among groups. Through focusing on my previous and current work on the Mandaya and the Mangguañgan of the eastern Davao cordillera, I want to show how intercultural and ethnic relationships work among the Mandaya, the Mangguañgan, and the coastal Bisayans, who are primarily Cebuano speakers (see Map 10.1). The interaction of all three societies requires development of two models of hierarchy, one of which fits the context of the interior or the center, while the other relates to the situation on the frontiers and coastal areas. However, before the empirical and theoretical analysis is undertaken, I would like to discuss some problems of the earlier ethnographic taxonomies of the Manobo, the Mangguañgan, and the Mandaya.

The Manobo are the largest tribe in eastern Mindanao, numbering from 35,000 to 40,000. Garvan (1931) provides a good account of the aboriginal Manobo at the turn of the nineteenth century. The Manobo are now found in western and eastern Bukidnon (Elkins n.d.:1–5), in all of the Agusan Valley except for the bordering areas of Misamis Oriental, in the coastal and mountain ranges of Surigao from Bislig north to the Cantilan River, and in northern Davao Province.

Rice represents the leading staple on the eastern side of the Pacific cordillera, while corn is the leading staple among those groups located in the Agusan Valley. The *kaingin* (swidden) pattern—the distance between settlement and *kaingin* and the use of secondary forests for agriculture—is similar to other regions within the Agusan drainage. *Kaingin* size ranges from two-fifths to 1 hectare per cultivator with each clearing utilizable for one growing season. Rice in the Agusan Basin has been abandoned, since crop yields have been reduced to eight to fifteen *cavan* of unhusked rice (*humay*) per year due to rat infestation and governmental prohibition against the cutting of primary forests for agriculture. Cassava, *camotes* (sweet potato), *ubi* (yam), *gabi* (taro), and bananas, along with various vegetables, are the remaining consumption crops, while *abaca* (hemp) is the sole cash crop. *Abaca* is harvested twice a year, with a yield of 500 kilos per hectare.

Manobo socioreligious life is described in detail by Garvan (1931:187–240) and in its basic features is similar to those of the Mandaya. The bilateral kindred is extended further laterally than among the less sedentary Mandaya. Most of the Manobo populations encountered in Agusan have had more Bisayan influence than the Davao tribes, except the Bagobo. Bisayan is spoken in the household, as well as being a *lingua franca*. Christian influence among the Manobo has extended itself as the aboriginal population became involved in the cash economy.

The Mangguañgan were estimated in earlier accounts as numbering from 10,000 to 20,000 (Garvan 1931:5) and inhabiting the upper Agusan, Manat, and

Mawab rivers. These areas are now occupied by the Bisayan and the Mansaka. Furthermore, if the Mangguañgan once lived in these areas, they are now difficult to trace.

In eastern Davao, the Mangguañgan groups occupy the upper slopes of the Pacific cordillera over an elevation of 3,200 feet. In upper Manay and Caraga, the Mangguañgan are located above the Mandaya populations and seldom descend into Mandaya territory. They are also found in southeast Compostela on the headwaters of the Caraga River.

Economically the Mangguañgan in upper Caraga and Manay subsist primarily on a marginal cultivation of tubers along with hunting, gathering, and fishing. Virgin forests are not cut and cleared for kaingin; only the undergrowth is cleared in small patches and planted in tubers. This marginal cultivation has not curtailed the nomadic movements of the Mangguañgan bands. A plot is cleared and planted to tubers, and the band returns to harvest the crop two or three months later. Their daily diet is mostly obtained by hunting and trapping of monkeys, lizards, wild pigs, and snakes; collecting of wild tubers, roots, and vegetables; and fishing for eel, shellfish, and fish. The band is a nomadic unit inhabiting a designated area and seldom trespassing into adjacent vicinities occupied by other bands.

Band composition from two observed cases consisted of fourteen and twenty people, both adults and children. The distance traveled each day by the band varies according to the amount of game and food encountered; however, the band elders recognize the abundance of certain game or food in various localities at specific times of the year. The routes each band follows are usually to and from areas of abundant food. Estimating the area each band occupies is difficult since the amount of time I spent with the bands was quite brief.

Spears, bow and arrows, bolos obtained from the Mandaya, bamboo knives, and fish poisons are the primary tools in obtaining a livelihood. Spears are made from wooden shafts that have been sharpened and are occasionally notched; however, iron points made from coins and tin cans are also utilized. Bolos, steel knives, and Western-style clothing are obtained from the remote Mandaya cultivators in exchange for beeswax and slaves. The *tubli* root is the most commonly used fish poison.

Garvan reported that the Mangguañgan were warlike and were continually involved in inter- and intratribal warfare. These conditions may have existed among wholly agricultural Mangguañgan populations in the Agusan Valley, but the upper Manay Mandaya have always recognized the Mangguañgan as a "weak people" from whom slaves were obtained. The slave trade is still common and constitutes the main line of Mandaya-Mangguañgan interaction.

The use of Cebuano-Bisayan is absent among the Mangguañgan; however, Mandaya and Mangguañgan are mutually intelligible. Mangguañgans who have had prolonged contact with the Mandaya have also learned Cebuano-Bisayan.

The coastal *conquistas* (Christianized Mandaya) designate the Mangguañgan as a branch of the Mandaya who inhabit the remote forest areas. Cole (1913:165) also lists the Mangwanga or Mangrangan (dwellers in the forests) as a branch of the Mandaya, although Garvan (1931:5-6) labels the Mangguañgan as culturally and racially different from the neighboring Mandaya. Among the non-Christian Mandaya of upper Caraga, the Mangguañgan are considered as a different and inferior people. When it was mentioned that the *conquistas* regard the Mangguañgan as Mandaya, the non-Christians were dumbfounded and horrified. From my observations I would also differentiate between the Mandaya and Mangguañgan on sociocultural lines, although linguistically they are only dialectically separated. Culturally the Mangguañgan are quite different and do not exhibit any of the political and religious aspects of Mandaya culture. The Mandaya are economically more stable and politically more integrated in comparison with the nomadic Mangguañgan bands.

The Mandaya are culturally quite representative of the Davao Gulf tribes, although variations are found in cultural emphasis in certain aspects. Previous studies of the Mandaya have enumerated a number of cultural-linguistic branches. Cole (1913:165) lists five synonyms under Mandaya or "inhabitants of the uplands": (1) *Mansaka*, or inhabitants of the mountain clearings, include Mandaya who formerly inhabited the interior areas and have recently moved into the coastal areas of Kingking, Mabini, and Lupon on the north and east side of Davao Gulf. (2) *Pagsupan*, or Mandaya living near the Tagum or Hijo rivers. (3) *Mangwanga* or *Mangrañgan* (dwellers in the forest) are Mandaya who live in the heavily forested mountains. (In the previous section, I designated the Mangguañgan as a distinctive group and thus would not include them as a branch of the Mandaya.) (4) *Managosan* or *Magosan* are Mandaya who live on the headwaters of the Agusan River now called the Managosan Valley. Most of the inhabitants are Mansaka/ Mandaya speakers, while the Mandaya speakers are only found in the more mountainous regions to the east. (5) *Divavaoan* is a division inhabiting a small district to the south and west of Compostela.

A survey by Svelmoe, Richert, and Thomas (1957:1-11) divided the Davao area into language groups and dialect areas. North and northeast Davao were included in the language group called Manobo, while the Davaweño language group "occupies the eastern Peninsula, the entire coastal area from about Cateel to below Digos, and inland from Digos to beyond the Cotabato border near Mt. Matutum" (Svelmoe, Richert, and Thomas 1957:9). Manobo includes four dialect areas: *Dibabaon-Mandaya*, found in Monkayo; *Mangguañgan*, only found in Gabi, northern Compostela; *Caraga Manobo*, spoken in Caraga; and *Governor Generoso Manobo* on the San Agustin Peninsula, akin to the Sarangani and Caburan Kulaman.

Davaweño includes the following ten dialect areas: (1) Davaweño or all coastal Moros; (2) Mansaka, found in Lupon, Pantukan, and the Maragosan valleys; (3) Mandayan, spoken in most of Compostela and closely related to Mansaka; (4) Mangaragan, spoken in upper Caraga; (5) Cateelano, spoken in upper Cateel and Baganga; (6) Manurigao, spoken in southwest Baganga and southeast Compostela, and appearing to be identical with Mangaragan; (7) Manay Mandaya, spoken in Manay; (8) Saug Mandaya, spoken in upper Tagum, Panabo, and Saug; (9) Samal (Isamal), spoken on Samal Island; and (10) Tagakaolo, spoken in areas west of Digos. The dialect differences noted by the Svelmoe, Richert, and Thomas survey for eastern Davao are present, but with modifications. Manay Mandaya, Manurigao, and Mangaragan are all mutually intelligible and have little dialect difference, at least not enough to divide them operationally into three dialect areas. This is due to the historical connection all three areas possessed as a common trade network with the Spanish at Caraga. These trade routes through upper Manay, upper Caraga, and southeast Compostela still exist and provide for areas of communication.

The dialect area called Caraga Manobo is also doubtful. Svelmoe, Richert, and Thomas (1957:9) state that the informant from whom a few words were obtained was not a native speaker from the area. Although Manobo dialectical influence has moved south from Agusan into northern Davao, its influence on the eastern Davao Coast is nearly absent except for northern Cateel, where Manobo-Mandaya groups are in closer interaction. To create a dialect based on a single nonnative informant does not seem warranted.

Division of the Mandaya by Cole and by Svelmoe, Richert, and Thomas tends to fragment the Mandaya into a number of cultural and linguistic subsets, creating a picture of tribal fragmentation. Although the Mandaya were never a single cohesive political entity, the tribal subdivisions have little bearing on the ethnographic picture as of the early 1960s through at least the late 1970s. The term *Mansaka* is only used for Mandaya who still maintain a traditional way of life in regard to the *kaingin*/shifting cultivation practices; seldom is it used as a designation for a particular locality.

Cole's *Pagsupan* or Saug Mandaya was always primarily Bisayan, though some Mandaya was spoken in the Tagum and Hijo River areas. However, these were Mandaya speakers from the east who were employed by local hemp plantations. Mandaya was not an indigenous language or dialect in that area.

As previously noted, the Mangguañgan (Cole's Mangwanga or Mangrañgan) should be designated as a distinctive ethnic entity due to differences in language and to economic and cultural features. These factors set them off from the more populous Mandaya who maintain connections with the Mangguañgan through trading and occasional forays to acquire slaves. They should not be grouped as a subbranch of the Mandaya.

The subdivision Divavaoan did not exist in the early 1960s when I worked in the Compostela area. Most of the aboriginal population of the Compostela area spoke Bisayan and, in reality, most of them were Manobo speakers who claimed that the Mandaya were found only in the cordillera area east of Compostela proper.

The ethnic divisions used by Cole, as well as by Svelmoe, Richert, and Thomas, are based on the assumption that proper names of localities are coterminous with a subethnic cultural designation. Within this framework, one can readily note how divisions are created through the fragmentation of a single language grouping or a single cultural entity into any locality where that language or society exists. Thus, Mandaya was only understood as a rubric, which embraced all these different peoples residing in different localities. In reality, the issue is to deal with a single complex, be it a society or a language, and to determine how local and regional variation evolves and exists, and to what extent truly significant differences emerge, if at all. By the 1960s, these differences, if they actually existed, were simply of little or no importance, and older informants could not shed any light on their actual historical existence.

However, the damage of heavy ethnic/linguistic fragmentation created by early ethnographers cannot be corrected. Over time, these small units appear on maps, become part of the ethnographic inventory of an area or a region, and are portrayed through the use of different colors on maps, and thus what started as a dubious division assumes a hard empirical reality. A glance at the upland populations in Vietnam and Laos immediately brings to mind a most minute fragmentation. This, in part, was the result of early ethnographic accounts that sought, wittingly or unwittingly, to divide large units into smaller, more compact entities that could be controlled politically by the colonial government. Political control was exercised either directly by the colonial power or indirectly through the creation of a native civil service, which held dual allegiance to the colonial system and to their own racial, linguistic, and cultural groups.

At the same time, we must understand the role capital penetration played in the Davao area. As has been noted elsewhere, the emergence of capitalism, be it in the form of the plantation system or through the employment of labor for the logging industry, has had a marked impact in creating a homogeneous labor force that was gradually denuded of its original cultural base. Throughout the late Spanish period (1850-1900) in the Davao Gulf area or on the Pacific coast, commercial impact was minimal, whereas the basic thrust of Spanish culture was expressed through a series of missions extending from northern Agusan and Surigao southward to eastern Davao and into the Davao Gulf region. The Christianization of native peoples occurred only among those who resided on the coast; thus, there was little impact on interior groups, in spite of Mandaya raids on coastal missions.

With the coming of the Americans in 1902, the shift from religious evangelization to commercial enterprise became all the more important. Plantations of hemp (*abaca*), fruit crops, and coconuts dotted the coastal regions of Davao Gulf, while economic penetration of the interior started through logging and small mining ventures. Along the Pacific coast, Mati was a prime locality for coconuts and hemp, while the coastal towns were usually devoted to coconuts since the coastal plain was not great enough to support large-scale enterprises. A plantation labor force was created from different native groups: mostly Bagobo on the western side of the gulf, Bila-an on the south gulf area, and coastal Mandaya in the northern and eastern sections of Davao Gulf. The consequent movement of tribal peoples to distant localities soon created an uprooted sense of community in which language and cultural differences among groups decreased and cultural homogenization started.

Also, by the 1910s and throughout the 1920s and 1930s, Bisayans—including Cebuanos, Boholanos, and Ilongos—moved into the area as plantation labor. Over time many of the settlers acquired land once owned by local native peoples. Furthermore, throughout the gulf region, intermarriage between Bisayans and local people started, which in turn added to the displacement of the population, the loss of traditional landholdings, and the blurring and obliteration of cultural and linguistic differences.

Cole, who worked in Davao Gulf and Mati in 1908 and 1910, might have noted "actual" differences in ethnic composition, but even by that time these differences were no longer significant. On the east coast and in the uplands, the cultural framework existed longer and only changed when commercial logging enterprises had pushed into the mountainous areas beyond the upper Agusan River drainage. As of the 1960s and 1970s, the contrast among Mandaya, Manobo, and Mangguañgan still existed on the east coast and interior mountains, but by then the Bisayans had moved into the coastal towns and settled as small-scale farmers of wet rice and coconuts.

To understand the dynamics that regulate the interethnic interaction among the Mangguañgan, Mandaya, and Bisayans requires an analysis of their perceptions of one another, and what kinds of structural relationships underwrite the total system of exchange.

Understanding Mandaya Interethnic Relations

Mandaya ethnography has been reported by Cole (1913), Garvan (1931), and Yengoyan (1964, 1965, 1966a, 1966b, 1966c, 1970a, 1970b, 1971, 1973, 1975a, 1975b, 1977, 1983, 1985); thus, only those aspects of the ethnography that relate to the issue of ethnicity will be discussed. On the eastern Davao cordillera, the Mandaya are shifting cultivators who occupy both the foothills, where they are involved in hemp production, and the interior uplands to an elevation of 3,200 feet in which a mixed system of dry rice cultivation and the farming of tubers is their basic mode of subsistence. Just as the subsistence base varies, the settlement pattern also covaries with economic demands. With dry rice cultivation, the settlement pattern consists primarily of a single household adjacent to cultivated fields. Households are moved as often as swiddens are relocated, which means virtually every year. Households are synonymous with the nuclear family; thus, each family unit is also the unit of production. Distances between swiddens vary from 0.5 to 2 miles. However, in the relocation and creation of new swiddens, households are situated so they are always in visual contact with at least one other household, either across the valley or on top of a range of hills. With the sedentary system based on hemp production, one finds the beginnings of settlement nucleation in which hamlets of five to eight households form a cluster and, in some cases, hamlets are also seminucleated in small villages.

The coastal areas are occupied by either the Bisayans, who migrated from Cebu, Leyte, and Bohol, or by the *conquistas*, who are the descendants of Christianized Mandaya. Interestingly, the *conquistas* do not consider themselves as Mandaya descendants, but as Bisayan, since they have been baptized. In fact, the Christian act of baptism is primarily a means by which one's identity changes from Mandaya to Bisayan; thus, spiritual rebirth has almost no bearing on the volitional act (see Yengoyan 1966a). Other influences are critical to understanding the means by which group identity comes about, but baptism among the Mandaya in the late nineteenth century is the key to understanding the emergence of the *conquistas*.

Most lowland populations either cultivate wet rice or plant coconuts on the relatively narrow plains, but over the past twenty years Bisayans and *conquistas* have moved into the foothills and started hemp cultivation. This penetration into the foothills and the usurpation of lands from the upland Mandaya have brought forth considerable conflict, since the Mandaya hold land by usufruct and their land is never surveyed and titled, while the Bisayans claim that title to land, and not actual possession, is the sole basis of ownership. Since the 1950s, the foothill Mandaya who have lost their land to Bisayan encroachments have either reverted to dry rice cultivation by moving into the forested interiors or have become part of the rural proletariat, working for minimal wages from Bisayan landlords on land they once possessed.

Coastal Bisayans not only maintain their hegemony through economic control, but most of the political and commercial power is in their hands. Although some of the larger shops in towns such as Mati, Manay, Cateel, and Baganga are owned by Chinese, the local Bisayans have either set up their own shops or have worked out financial arrangements with the Chinese in which the Chinese are "upfront" with their activities, while the Bisayans have invested in them and/or have offered them protection. Most local political positions are held by Bisayans, and in a number of cases two or three intermarrying families have developed a web of mutual interest and have virtually sealed off a town from outside intrusion. Because the coastal highway beyond Mati has never been completed and air service is unavailable, shipping is the only means by which outside contact is maintained.

As noted earlier, the Mangguañgan are located in the densely forested interiors where a seminomadic life-style is maintained through collecting, hunting, and trading of forest products with the Mandaya. As of ten years ago, almost no territorial conflict occurred between the Mandaya and the Mangguañgan, since Mandaya swiddens were seldom located at an elevation of more than 3,200 feet and did not impinge on the Mangguañgan.

However, the Mandaya have always raided the interior for "slaves," especially for young Mangguañgan females, who were then raised as domestic servants and later married among the Mandaya. The Mandaya claim that this is an old practice and remark that the Mangguañgan are a weak people since they do not cultivate rice though they know its value. Although the Mandaya refer to their slaves as *posaka*, and in some cases set them apart from normal domestic and religious relations, this seldom extends beyond one generation. The cultural and social hiatus that we attribute to slavery is practically absent, since most young Mangguañgan slaves maintain contact with their natal family.

Within this context of cultural triangulation, the Mandaya are in the middle since there is seldom any direct contact among the Mangguañgan and the coastal Bisayans and *conquistas*. The Mandaya perceive the Bisayans as either land-grabbers or potential land-grabbers who have used the political and administrative structure for their own benefit. Thus, the question of tax collection in which the Bisayans irregularly request that land taxes be paid simply increases the tension on both sides. The conflict is further provoked to higher degrees of tension when the Mandaya lost their land but decided to remain as laborers. Violence and homicide occur sporadically; the most common cause is usually the Mandayan's inability to maintain their landholdings against Bisayan encroachment.

Mandaya who think and act like Bisayans (i.e., by converting to the Catholic Church, wearing Western-type apparel, cutting their hair, and taking Christian surnames) desire to maintain their commercial transactions for selling hemp and acquiring credit with the Chinese merchants in the coastal towns. The Chinese usually deal in straight commercial transactions without the derogatory perceptions that characterize Bisayan-Mandaya relations.

Models of Structural Interaction and Politics of Ethnicity

The expression of domination and hegemony of either one social class over another, or of one society and culture over another, is normally based on how the ruling group, class, or society develops and maintains its control of resources, of services and labor (of which slavery is an example), or of symbols representative of control. Most studies of ethnicity have stressed the idea that wherever two groups compete for the same resources, a form of domination may start; yet in many societies, competing classes might read different meanings in one another's activities and if the differences are clear and stabilized, intergroup or interclass conflict might be minimized. Foster's chapter in this volume shows how the space created between the Thai and the Mon allows maintenance of commercial activities. Interaction still continues despite the Thai's inherent feeling that commerce is evil, though necessary. The potential conflict between Thai and Mon is subverted through the different interpretations of actions, symbols, and meanings that each group has of the other.

However, conflict occurs when two spatially juxtaposed groups must compete for the same resources and the same labor supply, but one group is able to maintain the differential by controlling symbols of domination. Such an example is provided in the Mandaya conflict with the Bisayans in the foothills where, through land expropriation, the Mandaya have been reduced to laborers. In turn, they realized that the political system and the local administrative structure, all of which symbolize the potential force of the state, are in the hands of the Bisayans. If political participation is closed to the Mandaya, the only other outlet they possess for acquiring and using certain symbols of domination lies in the religious realm. Thus religious conversion, either to Catholicism or to any Protestant denomination, is an important means of acquiring access to calling oneself a Bisayan. Usually most Mandaya who undergo conversion start and end with the baptismal rites. However, the local Catholic parish priests in many of the coastal towns will insist that the new converts return to the parish for additional religious instructions. But the Protestants are more lax on this matter, partly because they simply do not have the local personnel and logistics to follow up with further instructions.

Beyond the religious changes, the convert usually takes a Christian personal name and surname, serves as a godparent to those who ask him or her, and in most cases wears European clothes. At the same time, he or she may possess certain attributes, making it all the more difficult to be called a Bisayan. Traditionally, both the male and female Mandaya possess an extensive pattern of tattoos; however, male tattooing is more intricate in design and also covered more parts of the body. Although tattooing has gradually decreased in intensity during the past twenty years, the practice still exists. When tattoos are visible and cannot be covered by Western-style clothes, one's origin of birth is readily recognized. The Bisayans also recognize this pattern among their Mandaya counterparts and will always transmit such knowledge to another Bisayan. Thus, if a male wears long-sleeve shirts, especially in public, the Bisayans claim that is a sure sign the individual is a Mandaya by birth. For many individuals, there is in fact no visible route for making the transition into a Bisayan life-style.

From the foregoing description of Bisayan and Mandaya interactions, it is difficult to assess whether the nature of the conflict is primarily ethnic inter-relationships or whether this particular case can be analyzed in relation to structured class relationships. Here cultural and linguistic differences can be interpreted as surface phenomena, which in turn disguise the internal linkages of domination that maintain the subservient position of the Mandaya. At the same time, one can consider this a total system of social relations that are overtly manifest along marked class lines.

One major problem in ethnicity studies is the determination of the interconnections between culture and cultural differences on one side, and the development of a class structure on the other. Two societies or groups that must compete for similar resources or ends might stress cultural differences at one level, though they are part of a single class structure. Along these lines of inquiry, one can determine the extent of domination in class-related structures to the extent that cultural differences are overtly stressed. Theoretically, the issue of class and ethnic relations is most difficult to understand, and even more so in Southeast Asia where the issue of group boundaries and demarcations is most difficult to establish. In fact, one might go as far as claiming that the search for boundaries might not be warranted due to the shifting nature of populations, the expansive and/or retracting basis of societies, and the ability to create, transmit, and remetaphorize symbols for the maintenance of cultural and ideological ends. Apart from such theoretical issues, the ability to empirically understand particular processes of intergroup interaction requires a means of analysis that can account for the structural linkages which might express the nature of Bisayan-Mandaya-Mangguañgan interactions within a total system of explanation.

Within the triangulation of the three societies, I will primarily devote my analysis to the Mandaya since it is the society I have studied in more detail. Furthermore, as stated earlier, the Mandaya are the middle link between the Bisayans and the Mangguañgan since there is no direct contact between the latter two societies. In this sense, it might not be proper to use the term triangulation, for what is being described is not a triangle but more like a linear polarity, with the Mandaya as the central unit. Since I have already provided a summary sketch of each society, our concern focuses on how to understand the totality of the structure within a single framework. Here my aim is to go beyond cultural and language differences and similarities and to elicit structural properties, which may account for overt aspects of economic and political relations.

Two concepts will be employed to understand the dynamics of the process. One is the contrast between the interior of Mandaya land and the exterior expressed by the frontier. The interior/exterior contrast is empirically verifiable as well as theoretically interesting as a means of delimiting the direction of changes. Another set of concepts, borrowed from Louis Dumont (1975, 1980,

1982), is the idea of hierarchy, which I will discuss in two forms: vertical hierarchy and the contrast between inclusion/exclusion. In the summary, the interrelationships between the interior/exterior and the two forms of hierarchy will be discussed, and how this type of explanation might have implications for other cases beyond the particular ethnographic examples.

The interior refers to how upland Mandaya society is structured. Whereas most lowland rice cultivating communities in the Philippines are characterized by a two-class system (Lynch 1984), one of the basic features of the Mandaya is a virtual absence of the contrast of "us" and "them." All Mandaya are upland rice cultivators, all suffer from a shortage of rice, and all families must revert to the consumption of root crops when rice is no longer available. Tubers are disliked and, in most cases, they are considered only fit for pigs. Each family continuously plants tubers for pig food; yet, they all realize that at a certain time after the harvest, they will be forced to consume root crops until the next rice harvest is available. At the same time, each family is fully aware of what households consume more rice per year in comparison with other households. The range of variation from late 1960 to 1962 indicated that those families with higher rice yields consumed tubers about 4 to 5 percent of their yearly intake, while those at the lower end of the scale consumed root crops within a range of 30 to 35 percent of the total food intake. Although this differential exists and the Mandaya are aware of such a contrast, the egalitarian nature of the social structure is not disrupted or verbally denied. Because land is a free good and open lands for cultivation still exist, each person still has access to land as a resource.

One of the major means in which rice consumption differentials are minimized is the institutional structure of gambling, which is widespread and as important as cockfighting. Gambling occurs on a small scale throughout the year, but the major gambling feasts occur during the immediate post-harvest period. Virtually all households are involved in the gambling of rice. Theoretically, the ideal culmination of these gambling feasts, which might last three or four days and nights amid heavy bouts of eating roasted pork and rice and drinking more than ample amounts of rice wine and sugarcane wine, is to gamble what rice one possesses until virtually the total community-produced rice is controlled by six to eight individuals. In some cases, the centralization of rice might be controlled by two or three individuals, especially if gambling feasts are extended to six or seven days. When the feasts end and virtually all locally produced rice is controlled by a few individuals, either male or female, a long process of redistribution ensues in which individuals who have lost their yearly crop are replenished with rice. The key to understanding the redistributive process is to establish what one has gambled and what one obtains in return. Those families who were at the lower ranges of production are normally supplied with more returned rice than they produced and, in turn, those families who had a good harvest may lose in the process of redistribution. No family has any say about what other families will obtain on the return, since those individuals who now control rice through gambling will establish how the return is to be made and in what quantities. What gambling does is equalize marked differences in production, thus allowing poorer families to consume rice over a number of months before reverting to root crops and vegetables. Gambling is a structured mechanism that minimizes possible class divisions; consequently, self-esteem is never lost. Since gambling is conducted without the profit motive, what does it mean? Basically, gambling with the ability to win or at least come up near the top of the finalists is a representation of the skills one possesses. A smart and shrewd gambler is respected for being able to make a bold and creative move and for being capable of long-range planning.

At the turn of the twentieth century, the Mandaya had a warrior class (*bagani*) and a council of elders that coordinated political decisions, mostly in warfare; however, since the 1920s warfare has virtually ceased. Overall, relatively few class markers exist among the Mandaya, and when they do, such as in the consumption of a valued item like rice, the possible distinctions are narrowed. Contrasts such as "us/them" exist, but overall the cultural ethos works toward maintaining a cohesive and coherent social structure where marked differences in life-style do not disrupt the overall exchange of resources and services.

To a certain extent, the interior as expressed by near absence of "us and them" also characterizes Mandaya and Mangguañgan relationships. The Mandaya perceive the Mangguañgan as different and weaker, but the idea of difference is accepted as a major feature that inherently demarcates people from one another. Furthermore, the differences are such that both the Mandaya and the Mangguañgan need each other; the social and economic ties are symbiotic, hostility is seldom expressed, and "slavery" is not cast as a hostile act.

In summary, the interior may be portrayed as a situation in which symbiotic and nonparasitic social relationships cement families with one another in a highly cohesive way, and this pattern is also extended to the Mangguañgan. Contrasts such as "us/them" might erupt and could possibly occur, but the system of exchange through gambling and social relationships works against the emergence and crystallization of hard social cleavages.

The exterior or the frontier is virtually based on the opposite of what has been described. With the introduction of commercial hemp production, land is no longer a free good; thus land, like labor and specialized technology, is an expression of capital that is utilized for production. The control and manipulation of resources, labor, and symbols by the Bisayans have produced a number of deep social and economic cleavages between the Bisayans and the Mandaya and have promoted a classic two-class structure similar to most rural lowland rice-producing areas in the Philippines. Through the loss of land to Bisayan hemp planters and loggers, the Mandaya on the frontier have been reduced to a rural proletariat with virtually no political power. Occasionally, a Mandaya might make the transition and acquire land, but the price of this shift almost requires a complete acceptance of Bisayan cultural symbols. As noted earlier, some Mandaya diagnostic features might curtail the ability to make this shift.

The emerging class structure is based on a number of features such as economic relationships, political power, social interaction, and verbal discourse, which work toward the creation and maintenance of social differences that cannot be bridged. To a certain extent, the stress on differences by the Bisayans is a highly conscious act occurring in a number of social activities in which the Bisayans and the Mandaya are juxtaposed. Even if the Mandaya attempt to approximate the actions and symbols of the dominant class, the Bisayans consciously acquire new symbols from the outside that are utilized to enhance social distinctions. One of the major acts of social distinction is the way in which the Bisayans become patrons of local religious festivals. Not only does the display of wealth enhance one's personal position, but it also ratifies one's impor-tance to the provincial political hierarchy and the priesthood in the towns. The distinction between "us and them" separates Mandaya from Bisayan, as well as create a contrast within the Bisayans where affluent planters consciously establish distance between themselves and small-scale planters or the landless Bisayans.

While economic and social distinctions are enhanced on the frontier as part of capital penetration, a hegemonic social structure also evolves in which political leadership is expressed through a system of patronage controlled by a few individuals. To call these leaders a rural "elite" is inappropriate, since this economic and political domination has occurred within the past twenty years. However, the local and national Philippine political structure and bureaucracy also operate to maintain certain individuals in power positions through which economic and political patronage is channeled from government institutions. The ability to funnel resources to certain individuals and groups, and not to others, means that what occurs as local level development is selectively controlled for future returns. Within this structure, the Mandaya have virtually no power over their own decisions. Not only are they without power, but they are culturally at a loss to understand what power means and how it is used for pursuing consciously sought ends. The class formation requires a basic set of assumptions and knowledge that all individuals share as givens, and it is along these lines that the Mandaya are virtually at a loss. One can imagine that by now, however, many realize what the game is and how it is played, although I suspect that it is too late to significantly alter the state of affairs.

In contrasting the interior to the exterior as two different forms of social structure, it is critical to understand these structural expressions within the context of Mandaya culture. The interior, where the contrast between "us and them" is minimally developed and where class divisions are absent, is considered by the Mandaya to be the focus of cultural expression. Here the lengthy rice rituals are enacted after every harvest, the role of shamanistic activity is still vital, and the great mythic and historic past is essentially conveyed through an oral tradition (Yengoyan 1985). The exterior, where the frontier blends into Bisayan society and where commercial agriculture starts, still maintains a semblance of Mandaya traditional culture, but many of the structural and symbolic expressions are now undergoing change and variation. The emergence of a two-class structure has created a significant departure in what Mandaya cultural life means for individuals who are working for Bisayan landlords.

Although the interior/exterior contrast illustrates the nature of different social relationships and the direction of change, the two expressions of social life are also related to two aspects of hierarchy. In Dumont's (1975, 1980, 1982) concept of hierarchy, the constitution of any society is not just a matter of social and cultural units, but how these units are interconnected with one another. The most intuitively obvious form of hierarchy is *vertical hierarchy*, such as is found in the caste system in India, a subject Dumont has explored in much detail. Here, each caste is understood by how it relates to those superordinate castes as well as those castes subordinate to it. However, to fully understand the process, the total system must be comprehended as it impinges on each and every caste group. Hierarchy in this model means ruling castes control resources and symbols that express their dominant position; thus, these resources and symbols are of value in establishing where each segment stands in relationship to one another and the whole. However, the whole is not only constituted and created by the castes and by their internal relationships. What Dumont argues is that the whole is essential and, in some ways, prior to the dynamics of each case. Value is the critical element in the animation of the structure into action.

A second form of hierarchy exists in societies based on *principles of exclusion and inclusion*. Exclusion/inclusion is expressed in the ways in which small groups relate to larger ones, and in turn how they all relate to the maximal unit, which might be a tribe. Principles of exclusion/inclusion are expressed both by social and spatial demarcations, such that one group is juxtaposed to, and contrasted with, another at one level of activity; yet at a higher level of classification, they are combined with another contrastive category. Hierarchy is expressed in the whole, which embraces all the internal contrasts; consequently, each segment is comprehended as a balanced set of oppositions. The Nuer are the best example of this type of hierarchy. Vertical ranking does not exist, but the hierarchy is based on the whole and how the whole relates to the internal workings of the parts. Here society is conceived as a set of concentric circles that embed one another until the widest limits of the system are approached. In Nuer, the tribe interconnects with other Nuer tribes; however, the system of exclusion and inclusion also embraces the Dinka and the Anuak.

Returning to the Mandaya, the contrast between the interior and the exterior (the frontier) as two different types of social structure within a single cultural framework permits us to empirically understand the kinds of processes that have evolved. However, the contrast might have even wider theoretical significance if the concept of hierarchy is utilized as a theoretical underpinning. By interpreting the two contrasts as variant expressions of hierarchy, we might be able to generate other models for explaining different ethnographic cases.

The interior of Mandaya land exemplifies hierarchy based on principles of exclusion/inclusion. Hierarchy is not vertically developed but is expressed in a number of concentric circles that emanate from the most minimal (i.e., the family/household swidden) through a number of wider social forms (the community, the old *bagani* domains), to the Mandaya of upper Manay and Caraga, and finally embracing all the interactions with the Mangguañgan. What is critical to grasp is how each segment maintains itself in relation to the next inclusive unit. Although none of the units dominates the other, each does require a structure that maintains links from the minimal to the maximal. In this case, the exchange of goods and services, and trade, overlap all segments and provide a cohesive unit such that any one particular unit can only be understood as embedded within the whole. Hierarchy here is an expression of two tendencies. One is that each segment is constituted and interconnected in a way that promotes a system of maintenance through exchange. On the other side, the dominance of the whole establishes the "borders" with the uplanders, and in turn provides the basis of the internal connections.

For the exterior or the frontier, the vertical structuring of the hierarchy is dominant. The Mandaya are a social class regulated by their relationships to those segments of society that control resources and services—those segments in the structure that create and utilize the symbols of power, which bring us back to value. Interclass relationships not only embed what happens with the Mandaya within the total social milieu but also provide a screen through which Mandayan social perceptions are generated and acted out. Like caste in India, the understanding of any particular social group, class, or caste must account for the totality of relationships of domination and hegemony that underwrite the social world. The difference between Indian castes and Mandayan class relations is that in India, caste is a cultural given that is not questioned, while among the Mandaya, class relations and the ensuing hegemonic relations are not only imposed from the outside, but there is no prior cultural given in Mandaya society that accounts for and explains how and why such relationships evolved. As noted earlier, the Mandaya simply do not possess a prior cultural text to explain and internalize a class structure that now embeds them on the exterior (the frontier).

Hierarchy as a theoretical framework permits us to comprehend the nature of economic activities. As capital penetration increases throughout the foothills and as hemp as a plantation crop becomes more widespread, the pattern described for the exterior will gradually expand at the expense of the interior. The shift from an egalitarian system to one of vertical hierarchy will occur if and when the pattern of upland Mandaya land use changes to commercial crops. Labor-intensive commercial crops require the ability of landlords to mobilize a labor pool that can be effectively deployed, especially where landholdings are not centralized. Hemp in the foothills requires a fair amount of labor-intensive activities that are only met through the appropriation of Mandaya labor. The implications of this process for the discussion of hierarchy are that vertical hierarchy situations will increase at the expense of systems based on exclusion/ inclusion. In this sense, I think the process is irreversible and one can speculate that there are relatively few cases in which the opposite occurs. At the same time, cases exist in Southeast Asia in which people move from the lowlands into the uplands as marginal shifting cultivators to escape the vertical hierarchy that embraces class relationships and hegemonic structures.

Conclusion

In attempting to understand Mandaya interactions with the Bisayans and the Mangguañgan, my concern is to develop a framework in which different expressions of relationship may be comprehended within a more or less single theoretical set of ideas. Since the Mandaya have different kinds of ethnic relations with the Bisayans and the Mangguañgan, recognizing these differences at one level of analysis is recast within a system of hierarchy. Yet the question remains, how do various expressions of ethnicity in Southeast Asia relate to different concepts of hierarchy? Although all social relationships between two or more groups require the analysis of who controls resources, labor and services, and the symbolic structure, are all such social relationships an expression of ethnicity? Here I would argue that ethnicity as developed by Barth and others is based on the principle that there are socially and culturally perceived differences between groups and societies. In some cases, these differences are primarily cultural and linguistic, the kinds of differences anthropologists have historically stressed. Furthermore, how a society tends to stress its own uniqueness and difference depend on the nature of the encounter. In some cases, the stress on cultural uniqueness occurs at the periphery or the frontier, in others it emerges at the center, and in some cases it occurs throughout the society as sporadic and discontinuous (see Chapter 9 by Rosaldo, this volume).

The essential quality is the perception of difference and the nature of social bonds that either minimize or maximize it, such as the contrast between the Chinese and host societies throughout Southeast Asia. The Mandaya in the foothills realize that they are at the mercy of the Bisayans, unless they return to

the interiors and subsist by upland dry rice cultivation. Ethnic ties between the
Mandaya and the Bisayans exist but are recast along class lines, a form of
domination that will eventually disenfranchise the Mandaya of their cultural and
linguistic heritage.

Hierarchy, both as a vertical structure and a model of exclusion/inclusion,
provides a basis for understanding ethnic interaction and change within a total
framework. In applying this framework to the Mandaya, the differential
processes by which the Mandaya relate to the Bisayans and the Mangguañgan
are empirically expressed in economic and social transactions. Ethnicity and
class relations are predictable from the hierarchy model, which not only
illuminates the kinds of contrast in social dynamics but also the direction of
socioeconomic change.

To what extent is this hierarchy model applicable to other areas in Southeast
Asia? In summary, I would like to discuss the implications for the Kachin, Shan,
and Burmese (Leach 1954). Leach's analysis of intergroup process is probably
the most significant work on the subject in Southeast Asian ethnography. One
might even argue that the anthropological interest in ethnic relations and
ethnicity in general owes its intellectual debt to Leach's analysis of upland
Burma.

Leach designates the Kachin, Shan, and Burmese along cultural and
linguistic lines, but the essential feature for understanding cultural interaction is
developed through the *gumsa/gumlao* model. Basically *gumsa* is a ranked society
of aristocracy with either factual or fictive genealogical depth, which the chiefs
invoke to justify their mythic claims to traditional founders. Furthermore, the
mythic and genealogical heritage justifies and supports their domination and
control of serfs who the *gumsa* lords claim have no connection to the ruling
classes. *Gumlao* social structure is expressed as an egalitarian social network in
that domination and class hegemony is absent and, in theory, each productive
household unit controls its own ends. There are no masters and no serfs. Leach
describes this as a form of anarchic republicanism.

Of interest for my argument is that in tracing the transformation occurring
between *gumsa* and *gumlao*, Leach notes that most *gumlao* communities convert
to *gumsa* since upland rice lands are in short supply and, when people can no
longer sustain themselves, they join *gumsa* villages by pledging allegiance and
obedience to the *gumsa* hereditary chiefs. In fact almost all the shifts seem to
occur from *gumlao* to *gumsa*, and not vice versa; though in theory this could
occur, the logistics would be diffi cult to maintain. Eventually, *gumlao* structures and communities are maintained through extracting tribute from traders
and merchants who are involved in the China-Burma trade. The transformation
is one way from *gumlao* to *gumsa*.

In relation to my development of two types of hierarchy, Leach's *gumsa*
structures would fit what I have presented as vertical hierarchy, a system of

domination and hegemonic control in which class relations provide the basis of social articulation. Although Leach does not consciously discuss class structure as part of the *gumsa* model, one can readily note that the system of control and extraction is based on *gumsa* chiefs believing that they are born to rule through their hereditary connections with founding ancestors, as expressed and conveyed in mythic and genealogical structures. This aspect of historical domination has yet to evolve among Bisayan landlords in the foothills and frontier areas, but is common in the Philippine lowlands where history is always cited to justify and ratify the status quo.

Gumlao structures based on egalitarian principles are similar to the upland Mandaya dry rice cultivators. Households are autonomously structured; however, they are connected with one another through a system of labor exchange and mutual assistance. The pattern of exclusion/inclusion forms an all-embracing structure that extends to the outer limits of the social world, like the Mangguañgan, but none of these structures is in a position to control or dominate the other.

Within these two types of hierarchy, the movement, if it does take place, would be from the exclusion/inclusion model to vertical hierarchy. This pattern of transformation is similar to the change from *gumlao* to *gumsa*; and in the Mandaya case, the inducements are primarily economic (i.e., through loss of land, shortage of land, or desire to change to a cash crop such as hemp). Unlike the unstable *gumlao* structures, the upland Mandaya exclusion/inclusion context is more stable, since land does exist and population pressure is not yet a problem.[1]

Acknowledgments

I wish to thank Dr. Arjun Appadurai for his assistance in clarifying some of the theoretical issues in his succinct appraisal of my form of argumentation. This paper was completed while I was a Fellow at the Center for Advanced Study in the Behavioral Sciences, Stanford, California. I am grateful for financial support provided by the National Science Foundation Grant BNS-8011494 and by the University of Michigan, Ann Arbor.

Note

1. As an epilogue, since 1975 most of the foothill areas of eastern Davao have experienced armed uprisings by NPA forces, which have been able to fight the Philippine army to a stalemate. Armed conflict, in which the Mandaya have played almost no role, has in effect, then, saved them from

further land appropriation. Since the safety of planters cannot be guaranteed by the military, many landlords abandoned their lands and either settled in the safety of coastal towns or left the area. Hemp cultivation has declined while logging operations have been suspended; consequently, inroads for capitalistic penetration have been temporarily denied.

References

Cole, F.C.
1913 *The Wild Tribes of Davao District, Mindanao.* Chicago: Field Museum of Natural History.

Dumont, Louis
1975 Preface by Louis Dumont to the French edition of *The Nuer.* Translated by Mary and James Douglas. In *Studies in Social Anthropology*, edited by J.H.M. Beattie and R.G. Lienhardt. Oxford: Clarendon Press. Pp. 328–342.
1980 *Homo Hierarchicus: An Essay on the Caste System.* Revised edition. Chicago: University of Chicago Press.
1982 On value. In *Proc. British Academy*, Vol. 66. Oxford: Oxford University Press. Pp. 207–241.

Elkins, Richard E.
n.d. *An Outline of the Kinship System of the Western Bukidnon Manobo.* Malaybalay: Summer Institute of Linguistics.

Garvan, J.M.
1931 *The Manobos of Mindanao.* Memoirs of the National Academy of Sciences, Vol. 23. Washington, D.C.: Government Printing Office.

Leach, Edmund
1954 *Political Systems of Highland Burma.* Cambridge: Harvard University Press.

Lynch, Frank
1984 Big and little people: Social class in the rural Philippines. In *Philippine Society and the Individual: Selected Essays of Frank Lynch, 1949–1976*, edited by Aram A. Yengoyan and Perla Q. Makil. Michigan Papers on South and Southeast Asia No. 24. Ann Arbor: University of Michigan Center for South and Southeast Asian Studies. Pp. 93–99.

Svelmoe, Gordon, Ernest Richert, and David Thomas
   1957 Mansaka Survey. Summer Institute of Linguistics, University of North Dakota.

Yengoyan, Aram A.
   1964 Environment, shifting cultivation, and social organization among the Mandaya of eastern Mindanao, Philippines. Ph.D. dissertation, University of Chicago.
   1965 Aspects of ecological succession among Mandaya populations in eastern Davao Province, Philippines. *Papers of the Michigan Academy of Science, Arts and Letters* 50:437–443.
   1966a Baptism and "Bisayanization" among the Mandaya of eastern Mindanao, Philippines. *Asian Studies* 4:324–327.
   1966b Marketing Networks and Economic Processes among the *Abaca* Cultivating Mandaya of Eastern Mindanao, Philippines. Report to the Agricultural Development Council, New York.
   1966c Marketing networks and economic processes among the *abaca* cultivating Mandaya of eastern Mindanao, Philippines. Reprinted from report and abstracted in *Selected Readings to Accompany Getting Agriculture Moving*, Vol. 2, edited by R.E. Borton. New York: ADC. Pp. 689–701.
   1970a Man and environment in the rural Philippines. *Philippine Sociological Review* 18:199–202.
   1970b Open networks and native formalism: The Mandaya and Pitjandjara cases. In *Marginal Natives: Anthropologists at Work*, edited by M. Freilich. New York: Harper and Row. Pp. 403–439.
   1971 The Philippines: The effects of cash cropping on Mandaya land tenure. In *Land Tenure in the Pacific*, edited by R. Crocombe. Melbourne: Oxford University Press. Pp. 362–376.
   1973 Kindreds and task groups in Mandaya social organization. *Ethnology* 12:163–177.
   1975a Introductory statement: Davao Gulf. In *Ethnic Groups of Insular Southeast Asia: Philippines and Formosa*, Vol. 2, edited by F.M. LeBar. New Haven, CT: Human Relations Area Files Press. Pp. 50–51.
   1975b Mandaya. In *Ethnic Groups of Insular Southeast Asia: Philippines and Formosa*, Vol. 2, edited by F.M. LeBar. New Haven, CT: Human Relations Area Files Press. Pp. 51–55.
   1977 Southeast Mindanao. In *Insular Southeast Asia: Ethnographic Studies, Section 4; Philippines*, Vol. 1, edited by F.M. LeBar. New Haven, CT: Human Relations Area Files Press. Pp. 79–116.

1983 Transvestitism and the ideology of gender: Southeast Asia and beyond. In *Feminist Re-Visions: What Has Been and Might Be*, edited by Vivian Patraka and Louise A. Tilly. Ann Arbor: University of Michigan, Women's Studies Program. Pp. 135–148.

1985 Memory, myth, and history: Traditional agriculture and structure in Mandaya society. In *Cultural Values and Human Ecology in Southeast Asia*, edited by Karl L. Hutterer, A. Terry Rambo, and George Lovelace. Ann Arbor: University of Michigan Center for South and Southeast Asian Studies, Paper No. 27. Pp. 157–176.

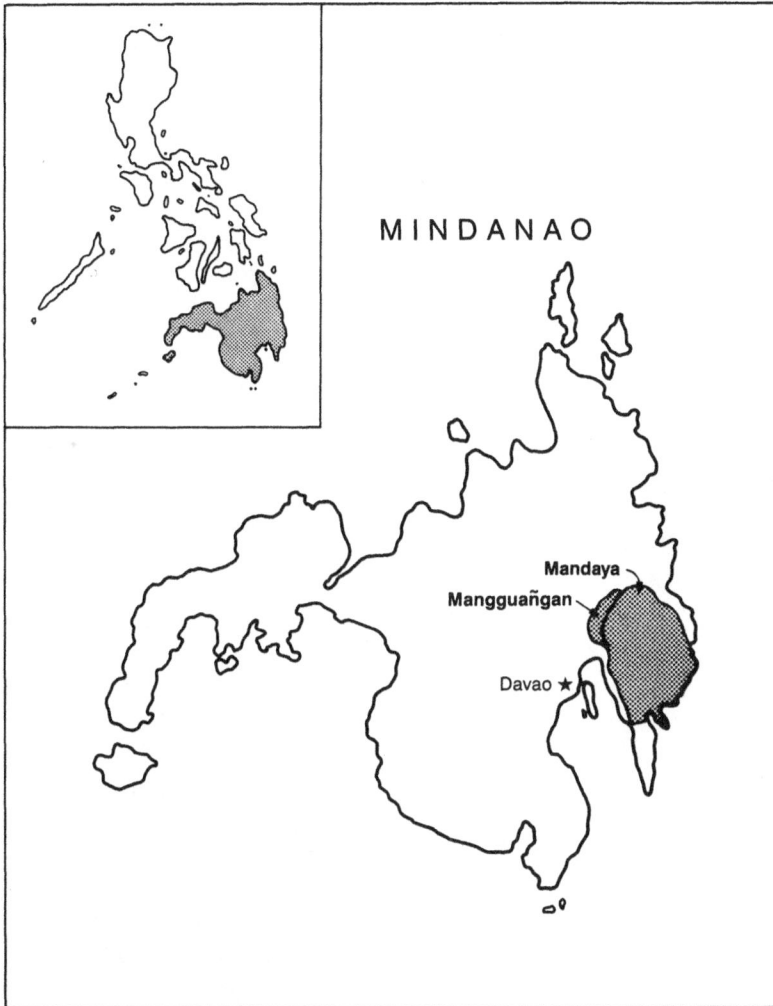

Map 10.1
Outline map of Mindanao indicating location of some ethnic
groups discussed in the text, excluding Bisayan migrants
(drawn from Yengoyan 1964)

CHAPTER 11

MATERIAL DEPENDENCE AND SYMBOLIC INDEPENDENCE: CONSTRUCTIONS OF MELAYU ETHNICITY IN ISLAND RIAU, INDONESIA

Vivienne Wee

Symbolic capital, a transformed and thereby disguised form of physical "economic" capital, produces its proper effect inasmuch, and only inasmuch, as it conceals the fact that it originates in "material" forms of capital which are also, in the last analysis, the source of its effects. (Bourdieu 1977:183)

In this chapter I follow Bourdieu (1977) in arguing for a comprehensive economics that would take into account both material and symbolic capital.[1] In the context of this argument, the symbolic economy is, to use Bourdieu's (1977:171) words, "the institutionally organized and guaranteed misrecognition" of the material economy. As the epigraph above indicates, it is through such misrecognition that the symbolic economy legitimates the material economy from which it originates. I propose to apply this argument to some ethnographic observations of certain island communities in the Indonesian province of Riau, where I have been doing research since 1979.[2]

The Material Economy of Island Riau

In official Indonesian terminology, the name "Riau" refers to the Sumatran province that includes the central part of the east coast, and the more than 3,200 small islands off the east coast, extending all the way to the South China Sea. In accordance with its geography, *Propinsi Riau* (Riau Province) is usually divided

197

into *Riau daratan* (inland Riau) and *Riau kepulauan* (island Riau). Administratively, island Riau forms *Kabupaten Kepulauan Riau* (District of the Riau Archipelago), which is thereby one of the six districts in the province[3] (see Map 11.1). In this chapter, however, I shall be focusing my attention on a small part of island Riau—to be specific, four *kepenghuluan* (headman-doms)[4] within *Kecamatan Bintan Selatan* (Subdistrict of Southern Bintan). These are the "headman-doms" of Penyengat, Pangkil, Bintan, and Karas. As Map 11.2 indicates, they are all located near Tanjungpinang. This is significant, because Tanjungpinang is the *ibukota* (capital city) of both the District of the Riau Archipelago and the Subdistrict of Southern Bintan. The communities that I shall be discussing are thus located in the political periphery of an administrative center at both the district and subdistrict levels.

According to the 1980 population registration, Riau Province has a total population of 2,174,266, with 423,495 of them in the District of the Riau Archipelago. My informants differentiate the population of island Riau into two categories—the *penduduk asli* (native inhabitants) and the *orang dari luar* (people from outside). My informants generally put themselves in the first category, although as we shall see below, who is native and who is from outside is by no means a clear-cut issue. Nevertheless, for first-generation migrants at least, it is possible to distinguish between who was born in island Riau and who was not. The blurring of the issue begins from the second generation onward.

Significantly, most of the first-generation migrants seem to be con-centrated in the major towns in the District, such as Tanjungpinang, Kijang, Tanjunguban, Tanjungbalai, and Dabo. In these towns, the migrants usually find employment in civil service, military service, construction industry, oil industry, granite quarries, bauxite mines, tin mines, clove plantations, rubber plantations, or various other businesses and service industries. In short, the people who move to island Riau usually do so on the basis of cash employment. Only a minority attempt to make a living through subsistence activities; interestingly, most of the people who try to do this are from Flores. Since cash employment is to be found mostly in towns and industrial centers, it is thus not surprising that the majority of migrants to island Riau are to be found there. As a result of this pattern of settlement, the duality between native and outsider can be mapped onto the duality of *kampung* (village) and *kota* (town). In this chapter, however, I am concerned only with the native inhabitants living in villages.

Although cash employment is more readily found in towns and industrial centers, the cash nexus is not limited to them; it most certainly extends to the villages beyond. What seems, from one perspective, to be the outward spread of the cash nexus from town to village is, from the opposite perspective, an economic continuum ranging from self-sufficiency within the village to other-dependence on the town. To illustrate this, let me compare three fishing techniques used by my informants—spear, line, and net.

Spear fishing is, to put it simply, the spearing of fish in clear, shallow water. A variety of spears and harpoons are used. For example, one is the *serampang* (trident), which has been described by Wilkinson (1959:1080) as follows:

> The two outer spikes of this fish-trident are barbed on the inside only; the central spike is shorter and barbed on both sides. All three meet in a common base, attached firmly to the wooden shaft by a ferrule or band.

The *gagang* (shaft) may also be made of bamboo. The *serampang* is used for the relatively smaller fishes. For larger prey such as dugong, the *tempuling* (single-pronged harpoon) is used. Although the single prong is also attached to a wooden or bamboo shaft, it is, in addition, affixed to a line and float. After the prey has been impaled, the shaft is detached from the prong and the line is played out; the reason is that unlike the smaller fishes, the larger prey can be pulled in only after it loses the tug-of-war.

To find the relatively smaller fishes, the spear fishers' boats are usually moored a short distance from the shore, just when the tide is going out. Their prey are the fishes being carried out by the tide. Such fishes are usually medium in size, small enough to be swept in and out by the tide, yet large enough to be seen by the naked eye and speared. As Sopher (1977:233) has pointed out, spear fishers

> have no means of taking some of the gregarious migratory fishes like the Engraulidae and Clupeidae, resembling the anchovies, sardines, herrings and shad of northern waters, and only take a few individuals from the schools of the larger fishes, like the mackerels and Spanish mackerels, or the flying-fish.... The oceanic pelagic kinds like skipjack, tunny and bonito, as well as porpoises,... appear to be quite disregarded by the sea nomads.... Their hunting techniques require them to concentrate on large, slow-moving surface fish, such as an individual Spanish mackerel, dorab or mullet and in the neighbourhood of coral reefs, the parrot fish, constituting a brilliantly coloured and delicately flavored prey, easily seen in the clear water near the face of the reef. In addition, sharks and rays, whose capture is always dangerous, are occasionally caught by the sea nomads, partly for food, but also for their commercial value.

Apart from these fishes, they also hunt turtle and dugong.

Although the spear fishers' catch is not without commercial value, and although they do occasionally sell it to obtain some cash, for the most part they subsist on what they catch. If they were to sell their catch regularly and then live on bought food, they would soon starve, for spear fishing is too inefficient a method for obtaining a sufficient quantity for sale. So if the spear fishers were to decide to live off the sale of their catch, they would have to change their fishing technique.

This is indeed the point: People who use spear fishing as their main technique do so not because they are ignorant of other methods of fishing, but because this method is the most suitable for their economy of subsistence. A few medium-sized fishes would be sufficient provision for a family for one day. Since fish is most certainly a perishable food that does not last beyond a few hours, particularly in the tropics, any other method of fishing that would obtain a greater quantity of fish would be counterproductive for an economy based on day-to-day subsistence.

In connection with this, Sopher (1977:234) has commented:

> Only small amounts of food are available to them from day to day and ... they have few means of preserving and storing an occasional large surplus. Thus, none of the sea nomad groups practise any of the methods of preserving fish by fermentation or marination which are common among a seaboard people of the Malaysian high cultures. An economic factor which is of significance here is the cost either of manufacturing salt, an essential ingredient, in terms of effort, time, and equipment, or of obtaining it in trade.... Smoking and drying of fish over a wood fire are hardly known in the western part of the archipelago, but are very common methods in Celebes and the Moluccas. In the absence of this method, and without the use of considerable amounts of salt, preservation of fish by simple means become difficult in the area of the Strait of Malacca, the east coast of Sumatra and the Pulau Tujuh, because of climatological reasons which make sun-drying often impracticable. The same climatological reasons, namely the continual raininess and heavy cloud cover, also make the production of salt from sea water a long and relatively expensive operation, particularly in the Strait of Malacca where the discharge of fresh water and salt is enough to reduce salinity considerably.

In my field observations, I have noticed that only net fishers practice fish preservation, usually by various means of salting. This is further illustration of the spear fishers' orientation toward a material economy of day-to-day subsistence, for they do not appear to be interested in preserving excess fish for future consumption.

Line fishing is carried out by dropping baited hooks on weighted lines into relatively deep water. Fishing rods are not used; the line is simply held in the bare hand. The kind of fish that is obtained by this method is the relatively large deep-water fish favored by Chinese buyers, especially restauranteurs. While line fishers also consume the fish they catch, the economy in which they operate is based not on day-to-day subsistence, but on trading fish for other goods. Since line fishing is carried out in relatively deep water, the fishers usually do not return to their home base after obtaining their catch, but instead take it straightaway to the market. Those who fish farther away from urban centers used to sell their catch to intermediaries known as *peraih*, who would wait for them at appointed meeting places on the sea at an appointed time, usually at dawn before the market in town opens. But this happens rarely now, partly because many of the line fishers have motor-powered boats that can speed them to the markets in time.

Of the three techniques under discussion, net fishing is the most capital-intensive. The construction and maintenance of nets require so much time and money that they have to be acquired ready-made. The prices of the nets vary depending on the kind of fish they are designed to catch. Prawning nets, for example, are so expensive that they are owned by the fish dealers and contracted out to individual fishers. The condition of the contract is that all the catch has to be sold at a fixed price to the particular fish dealer who owns the net. The contract is enforced by the existence of an agreement among the dealers themselves, such that if a contracted net fisherman were to sell his catch to anyone other than the owner of his nets, he would subsequently be boycotted by all the other dealers in the market. Therefore, the contracted net fishermen are, in effect, employees of the fish dealers, drawing their wages, like salesmen, on a commission basis, depending on the amount of their catch. In the case of nets for catching *ikan tamban* (herring), however, the nets have to be bought outright by the fishers. Both the nets for prawns and herring last only for a few months— perhaps six to eight months. Thus, a periodical capital outlay is necessary on the part of both fisher and fish dealer.

There are important differences between spear and line fishing on the one hand, and net fishing on the other. Both the spear and the line fishers own their means of production, whereas the net fishers do so only partially. Although the spear fishers buy the metal spearhead and prong they use for their trident and harpoon, as well as the nylon line they use for the latter, they make the wooden or bamboo shaft themselves, including the float for the harpoon. Their equipment is thus a mixture of self-sufficiency and other-dependence, with a potential for complete self-sufficiency, since the most rudimentary fish spear would simply be a sharpened wooden shaft. In contrast, the equipment of the line fishers—

that is, nylon line and metal hook—has to be entirely bought. But this is relatively cheap, and once a stock of lines and hooks has been acquired, the line
fishers are self-sufficient. In contrast, the net fishers are not even able to own
their means of production completely and are dependent on the fish dealers to
whom they are under contract.

Interestingly, among my informants, only the net fishers refer to themselves
as *nelayan* (fishers); the line fishers refer to themselves as *orang berniaga* (people
who trade), whereas the spear fishers are referred to both by themselves and by
others as *orang laut* (sea people). Both the *nelayan* and the *orang berniaga* look
down upon the orang laut as being primitive: the *orang laut* are said to be "dirty"
(*kotor*) because they "eat anything that is available" (*makan sebarang*). Subsistence on natural resources is thus despised by those who are involved in the
exchange economy. This derogatory attitude is known by the spear fishers,
prompting one of them to say to me:

> So what is wrong being an *orang laut*? Doesn't everyone in Riau
> live by the sea? Doesn't everyone depend on the sea? So isn't
> everyone an *orang laut* [a sea person]?

From the perspective of the *nelayan* and the *orang berniaga*, however, they
themselves are not *orang laut* because although they are fishers, they also eat rice
as their staple. Not only is rice non-aquatic, but it is not grown at all in the Riau
and Lingga archipelagoes. In recent years, a rice-planting experiment has been
conducted by the Indonesian government in one of the Tujuh islands—Natuna.
But that seems to be the only incident of rice-growing in the island world of
Riau, Lingga, and Tujuh. There is indeed rice grown on Sumatra, even in the
mainland area of Riau Province, such as Indragiri and Kampar. But the
province as a whole is not self-sufficient in rice; it has been estimated that it
needs to import 60,000 to 100,000 tons per year (Hendra Esmara
1975:31). Certainly, for the inhabitants of the island territory, rice is necessarily
a trade item. Anyone who eats rice as a daily food must therefore be involved in
the exchange economy in one way or another. The *nelayan* and the *orang
berniaga* are regular rice-eaters, whereas the *orang laut* subsist on the marine
products that they obtain themselves, eating rice only as an occasional exotic
food.

As I stated earlier, there is an economic duality of self-subsistence versus
other-dependence. The latter is aggravated particularly by a national economy
that is regulated by discretionary controls. As Booth and McCawley (1981:17-
18) have pointed out:

> Prices are fixed by regulation for a number of important items
> such as rice, fertilizers and kerosene and the government

attempts to inhibit the appearance of black markets by directly controlling supply. Licensing arrangements are common for a wide range of imported goods, and bank credit tends to be rationed according to the rather arbitrary imposition of credit ceilings rather than through interest rates. . . . Discretionary controls of this sort have been found to be unsatisfactory in many countries, and the experience in Indonesia does little to suggest that there is more justification for using them there than elsewhere. In Indonesia they have often been rather ineffective in controlling market forces and the arbitrary nature of the controls which allows administrators room for discretion in applying the regulations encourages corruption.

To enter the cash nexus in the context of such a national economy is thus to be subjected to numerous forces beyond one's control. The *nelayan* and the *orang berniaga* in Riau who enter the cash nexus can do so as no more than ordinary citizens. Therefore, they can have no control over the value of the *rupiah*; the prices of rice, kerosene, and imported goods; or access to bank credit. Even as the sellers of fish, they have no control over the fish market, being dependent on the fish dealers for marketing the fish to the ultimate consumers in Tanjungpinang and Singapore. In the case of the net fishers, they are even dependent on the dealers for their very means of production—namely, the prawning nets.

Such being the case, people who can subsist independently of the cash economy may be regarded as materially self-sufficient, whereas those who are drawn into the cash nexus may be regarded as materially other-dependent. The three fishing techniques discussed earlier illustrate the continuum between the two poles of this economic duality. The spear-fishing *orang laut* are almost entirely self-subsistent, while the net-fishing *nelayan* are almost entirely other-dependent. The line-fishing *orang berniaga* are in-between. On the one hand, like the spear fishers, they use a labor-intensive technique. On the other hand, like the net fishers, they are involved in the exchange economy. Since the number of fishes that one can catch with a line is rather limited, this constrains the profit that can be derived from line fishing. And indeed, the line fishing communities usually do not trade in fish alone, but are also involved in the trade of other products they can collect, such as sea-slug, shell fish, and seaweed.

The division of labor differs in these three fishing techniques. In spear fishing, both men and women participate in the activity of getting their daily food. For example, a married couple may set off with one partner rowing and the other spearing. Or if a whole family is dwelling on a boat, then spear fishing would be a family activity. In contrast, line fishing is done mostly by men, although occasionally women may accompany them, especially if the catch is for home consumption. The collection of sea-slugs, shell fish, and seaweed is usually done by women in the line-fishing communities. There is a sexual division of labor in the collection of natural products for trade, which contrasts with spear fishing on the one hand and net fishing on the other. Whereas spear fishing is carried out as a family activity in which both male and female may participate, net fishing is an exclusively male activity. Thus it is only the net fishers that we may generally speak of as fisher*men*. In the net-fishing communities, women do not go to sea with men nor do they collect natural products for trade. Widows are the only women who are supposed to earn money for their own living—for example, by selling cooked food, mending nets, or weaving thatch. The economic duality of self-subsistence versus other-dependence is thus expressed also through gender: In the former mode, there is economic commonality for both sexes, whereas in the latter mode, there is economic differentiation based on gender. The other-dependence of the net-fishing community as a whole is symbolized by the other-dependence of the women who are domesticated right out of economic production, becoming totally dependent on men. The other-dependence of men, however, is disguised by their social role as sole economic providers.

In the four "headman-doms" under discussion, spear fishing and line fishing may be found in the "headman-dom" of Karas, while net fishing may be found in the "headman-doms" of Pangkil and Penyengat. Interestingly, the people in the "headman-dom" of Bintan do not fish at all. The Bintan people describe themselves as *orang darat* (land people) or *petani* (cultivators). The people we are discussing thus include *orang laut, orang berniaga*, and *nelayan*, as well as *orang darat/petani*. Given the socio-economic implications of the fishing techniques discussed earlier, it is clear that these different communities are involved in the cash nexus in varying degrees.

With regards to the cultivators of Bintan, their main crop is rubber, plus some coffee, cassava, and various kinds of fruit. They grow no rice, even though they, like the line fishers and the net fishers, also eat it as a staple. In order to obtain the cash with which to buy rice, they tap and sell rubber latex. Therefore, the cultivators of Bintan are as other-dependent on the town as are the net fishers of Pangkil, despite the possibility of self-sufficiency through cultivation of more food crops.

Such being the case, we can locate the four "headman-doms" in terms of a material continuum ranging between self-sufficiency and other-dependence (see Figure 11.1). The respective location of the different communities on this continuum is in accordance with their relative involvement in the cash nexus.

As mentioned earlier, both spear fishing and line fishing may be found in the "headman-dom" of Karas. Among my informants in this "headman-dom," those who are spear fishers live on Nanga and Teluk Nipah, located off the east coast of Galang Baru (see Map 11.2). These spear fishers describe themselves as *orang laut yang diam rumah* (sea people who live in houses), in contradistinction to their boat-dwelling kin who are described as *orang laut yang diam sampan* (sea people who live in boats), or more succinctly, *orang sampan* (boat people). Both use spear fishing as their main technique; both thus tend toward the self-sufficient end of the material continuum. However, as I have pointed out earlier, even spear fishers do occasionally sell their catch for cash. So to say that the spear fishers tend toward material self-sufficiency does not mean that they have no access to trade goods whatsoever.

Indeed they do obtain trade goods, but only on a casual basis. For example, they may wish to acquire, say, a radio; for this, they need cash that may be raised in a number of ways—through selling marine products such as fish, crab, dugong, sea-slug, and seaweed, and through hiring themselves out as temporary labor for cutting wood. When sufficient cash has been accumulated, the radio would be bought, and that would be the end of the cash-earning activity. They would then revert to their self-subsistence, until such time when another trade good is desired.

In terms of their involvement in the cash nexus, there is a discernible difference between the house-dwelling spear fishers and their boat-dwelling kin. The former tend to place a higher value on the accumulation of trade goods, even if these are obtained only on a casual basis. In this context, it is significant that on Nanga and Teluk Nipah there are no other fishers using other fishing techniques; there are instead two merchants, one on each island, who are involved in sundry trading activities. These merchants buy whatever salable products that may be offered to them by the inhabitants of the area, and sell in turn whatever products that may be in demand. The products that are bought and sold cover a very wide range indeed, including marine products, forest products, and consumer goods. Thus, they may buy, for example, sea-slug and seaweed from an individual and, in exchange, sell a pair of Western-style leather shoes to the same person. For the house-dwelling spear fishers to live in proximity to such merchants means that they have easy access to the cash nexus, if and when they choose to enter it. Indeed it would seem that this easy access to the cash nexus is still a major incentive for shifting from a boat-dwelling to a house-dwelling way of life. I would even go so far as to say that in island Riau, house-dwelling spear fishers are almost always clustered near some such sundry merchant.

In contrast to the spear fishers, the line fishers are much more committed to the cash nexus. Among my informants in the "headman-dom" of Karas, those who are line fishers live on Karas island (see Map 11.2). These are the people who consider themselves *orang berniaga* (people who trade), and for whom fish constitutes only one resource among many. They are very active traders, whose trade route extends from Singapore to Jambi in Sumatra. Practically every able-bodied man on Karas island goes off on regular trading expeditions. For someone who wishes to start with no capital, one way of doing so would be to borrow a line and hook, go fishing, sell the fish in town, then with the cash that is acquired, collect some goods that would sell at the Barter Trading Station in Pasir Panjang, Singapore. The goods taken to Singapore may include sea-slug, seaweed, copra, rubber latex, and wood. The goods brought back home may include textiles, crockery, and electronic goods. Those who trade in Jambi are not beginners, but people who have already accumulated a substantial capital. For the Jambi trade, they go to Sungai Jodoh in Batam where there is a center for traders who have just returned from Singapore to off-load their goods. Although prices there would be more expensive than they are in Singapore, they are still cheaper than in Jambi. One of my Karas informants prospered so much from the Sungai Jodoh-Jambi trade that he was able to go on the pilgrimage to Mecca.

Perhaps because the Karas people are not dependent upon the conversion of a single product, such as fish or rubber, into cash, they are more flexible in their attitude toward available natural resources. They are not adverse to going to the forest to gather wood, or to attend to their fruit trees, or to collect marine products. In short, they use just about any means available to keep their petty entrepreneurship going. Nevertheless they are absorbed into the exchange economy, and like the people of Penyengat, Pangkil, and Bintan, most of their daily food is bought.

Unlike the house-dwelling spear fishers who have access to the cash nexus only by living near an individual sundry merchant, the line fishers (people who trade) participate in the cash nexus by traveling to urban centers. However, like the former, the latter's involvement in trade is still very much a matter of choice. They can still choose when and where they want to go, and what they wish to trade. Their material economy is still very mixed and therefore relatively unspecialized.

In contrast to this situation, the people in the "headman-doms" of Pangkil and Bintan have a much more specialized economy. To discuss Pangkil first, net fishing is the main source of income for the people, supplemented for some by the sale of coconuts. Almost every able-bodied man is contracted to a fish dealer in Tanjungpinang, to whom he is obliged to sell his catch. Despite this contractual burden, the Pangkil people are fairly prosperous, the sea being sufficiently full of fish that they are able to bring in a good catch most days. With the money they obtain from selling their fish, they buy commodities such as rice, vegetables, fruits, sugar, and salt. While it is true that they cannot themselves produce rice, sugar, and salt, it is nevertheless possible for them to grow vegetables and fruits. Since Pangkil is less densely populated than Penyengat and has its population concentrated only at the northern and southern ends of the island, there is a lot of land available for cultivation. But strangely enough, apart from some coconut groves, nobody makes any effort at growing vegetables or fruits. The only people who do so on the island seem to be a lone Chinese family who live by themselves on the eastern coast, somewhere between the two settlements; they grow eggplant for sale in Tanjungpinang. The reason my informants gave me for their lack of interest in plant cultivation was as follows: To make a *kebun* (garden), one must attend to it by living nearby. This would, however, result in one feeling *sunyi* (desolately lonely), since one would be separated from the neighbors by one's "garden." In order to be *ramai* (joyously crowded), one must live in proximity to one's neighbors; hence no "garden" is possible for lack of space.

Apart from this stated reason, there is another ecological one: When people on Pangkil wish to clear some land to build a house, they clear the land completely that the soil is left bare, uncovered by even a single blade of grass. After the rains fall a few times, the topsoil is washed away and all that is left is sand and gravel. The new house that is built thus stands in a sandy yard devoid of all vegetation. "*Pasir bersih*" (Sand is clean), my informants say. It is true that compared to dark soil, sand is whiter and drains better; in that sense it is cleaner. Evidently, where soil is concerned, my Pangkil informants prefer cleanliness to fertility.

This hostility toward vegetation is also discernible in their attitude toward the forest, which is said to be "dirty" (*kotor*) and full of "spirits" (*hantu*). Some secondary forest and scrub separate the two settlements on Pangkil—that is, Tanjung Budus in the north and Tajur Kait in the south. First of all, there is little social interaction between the two settlements. Second, if people want to get from one place to the other, they prefer to use their boats, rather than walk through the forest and scrub, even though that is only a half-hour's journey.

The only crop that the Pangkil people grow with any serious intent is coconut, which, as it happens, thrives even on sandy soil. Moreover, as mentioned earlier, the coconuts are sold. So although the owner of the coconut trees may keep a certain portion of the fruit for personal consumption, the rest constitutes a sales item to be converted into cash. The focus is thus on cash and not subsistence.

An interesting comparison is Bintan, where people who live by a river, not too far from the sea, do no fishing whatsoever. As mentioned earlier, my Bintan informants describe themselves as *orang darat* (land people). Evidently, they take this description seriously that they would only buy fish and do no fishing themselves. Even though the fish is very often stale by the time they buy it, they nevertheless persist in their abstinence from fishing. Their main source of income is rubber-tapping. Although the rubber trees are owned by a few of the wealthier individuals in the community, everyone does tapping. A few desultory fruit trees are grown, such as papaya, jack fruit, and mango, but no vegetables are cultivated. Like the Pangkil net fishers, the Bintan rubber-tappers use the money they earn to buy food instead of trying to obtain their food directly from available natural resources. Sometimes, however, they will cook the unripe fruit from their trees as a kind of vegetable.

There is thus a symbiotic relationship between the rural hinterland of Pangkil and Bintan on the one hand, and the urban center of Tanjungpinang on the other. In their single-minded devotion to a cash income, the people of Pangkil and Bintan may perhaps be regarded as peasants who supply raw materials to the urban center and who in turn obtain their daily necessities from there.

But if Pangkil and Bintan are symbiotically related to Tanjungpinang, Penyengat may be considered one of its suburbs. The proximity of Penyengat island to Tanjungpinang allows a substantial proportion of the island's population to work as *pegawai* (salaried employees) in town—mostly as minor civil servants and as employees of private companies. Only a small sector of the population derives an income from net fishing. In addition, a supplementary source of income for those with boats is to provide a ferry service between Penyengat and Tanjungpinang. This regular ferry service means that the people of Penyengat can get to Tanjungpinang very easily, not only for work but also for shopping, schooling, and entertainment. In short, the Penyengat population are effectively urbanized and almost fully integrated into the material economy of Tanjungpinang. The small number of coconut trees on the island are owned by a few individuals who sell the fruit not to the other islanders, but at the Tanjungpinang market. So of the four communities that we are discussing, the Penyengat community is the most other-dependent.

As the discussion indicates, important roles are played by individual merchants who service the rural population, as well as by those who are in the trade and employment sector of urban centers such as Tanjungpinang, Jambi, and Sungai Jodoh in Batam. It is through entering into a relationship with one or several of these people that one may enter the cash nexus. Such a relationship may be based on exchange, contract, or employment. The existence of these merchants and employers is thus an important fact in the everyday lives of my informants.

Apart from these, there is another category of people whose existence is also an important everyday fact. These are the people referred to as *pegawai pemerintah* (government employees), whose duties impinge upon the movement of the material economy, particularly if they are employed in the police force, the army, the navy, or Customs and Excise. The last-mentioned is perhaps the most relevant of all. Every fishing boat that goes to Tanjungpinang, even on a daily basis, must get official clearance to enter and leave the port. Every trading vessel that enters and leaves Indonesia must similarly get official clearance to do so. Since to obtain such official clearance one usually has to pay heavy fees for the licenses and heavy duties on the goods imported, many people try to evade this bureaucratic hurdle. Smuggling thus occurs as a consequence of tax evasion, the taxes being disproportionately onerous.[5] For my informants, the center from which government control emanates is Tanjungpinang. So both commercially and administratively, the importance of Tanjungpinang in their everyday lives is not to be underestimated.

The Symbolic Economy of *Melayu-ness*

The various communities we are discussing differ significantly from each other in material terms. Not only that, they have almost nothing to do with one another in the context of their material economy. They all trade with Tanjungpinang but not with each other. The people of Pangkil and Bintan do not trade fish for rubber. Nobody goes to Karas island to buy the goods of the traders there. It is the Karas traders themselves who take their goods into Tanjungpinang, Jambi, and elsewhere to sell. The bosses of the salaried employees of Penyengat are in Tanjungpinang. Even the spear fishing *orang laut* go to Tanjungpinang occasionally—for example, when they want to sell a dugong that they have caught, or when they want to buy some article not available from the sundry merchant near them.

If we were to consider these communities solely in terms of their material economy, we would perhaps picture them as shown in Figure 11.2. In such a picture, Tanjungpinang is the urban center to which the various places in the periphery are related, each separately and in varying degrees of closeness. This picture would seem to be an appropriate one in the context of transport logistics. For example, I myself, as someone without a boat, found it easier to go from one island to another by hitching rides to and from Tanjungpinang, rather than trying to catch a ride with someone going from, say, Pangkil to Karas island. The former route—that is, to and from Tanjungpinang—was an everyday occurrence. Fishing boats traveled to Tanjungpinang daily; but people on one island visited those on another island only on special occasions, such as weddings and funerals.

This gives us a clue to the nature of relations among the various island communities: These are not material but symbolic relations, having to do less with fishing and more with weddings and funerals. Indeed, it is only in terms of a symbolic economy that we may talk of these communities as being interconnected. In doing so, I follow the usages of my informants in these communities. In their perceptions of Self and Other, they were concerned with relating themselves not to Tanjungpinang, but rather to one another in the different island communities. This relationship is couched in terms of being *Melayu*, such that one could be "pure" or "impure," "indigenous" or "foreign" in one's *Melayu-ness*. Thus in the everyday discourse of the people living in these island communities, we discover a symbolic economy that is markedly different from their material economy.

The symbolic economy of *Melayu-ness* is expressed as a hierarchy of *derajat* "ranks":

| "aristocrat" | *tengku/raja* |
| | *tuan said* |

"commoner"	*encik datuk*
	encik keturunan
	orang biasa
	keturunan Bintan

| "serf" | *hamba raja* |
| | *hamba orang* |

As shown above, each rank is referred to by a specific term such as *tengku, raja,* and *tuan said*. One inherits a rank on this hierarchy through patrilineal transmission. Therefore, to be a *tengku* one must be descended from a patriline of *tengkus*; this applies even to the lower ranks, so that female hypogamy would not disturb the rule of patrilineal transmission. That is to say, a high-born woman and a low-born man would produce a low-born child.

If rank is regarded as important, and if rank must necessarily be inherited from one's patrilineal ancestors, then it follows that to have such a hierarchy is to interpret time in terms derived from the historical past. Indeed, my informants are themselves very aware of the significance of the past. Some of them even have written genealogies prominently displayed in the living rooms of their homes to show who they are on the basis of where they have come from. The past that is relevant to them is referred to as *zaman sultan* (the era of the sultan), beginning from the reign of Sultan Sulaiman (1721–60) and ending with the reign of Sultan Abdul Rahman Muazzam Syah (1885–1911).[6]

There is thus a quality of given-ness about this hierarchy. What is derived from a given past is regarded not as something that is being constructed, but as something that is simply so. A symbolic economy that is an inheritance from the past is seen as independent of the present—that is, independent of Tanjungpinang, the fish dealers, the sundry merchants, the Barter Trading Station in Singapore, the Indonesian bureaucrats, and all the other significant phenomena of the everyday present. The symbolic economy is the-past-in-the-present, a bracketed unchanging part of ongoing time. As Bourdieu (1977:163) has remarked:

> the temporal forms or the spatial structures structure not only
> the group's representation of the world but the group itself,
> which orders itself in accordance with this representation.

Indeed, this temporal form—this hierarchy derived from the past—is the template on which my informants pattern their *Melayu-ness*.

The significance of *Melayu-ness* has to do with the perceptions and cross-perceptions of Self and Other. It has to do with the location of one's self in relation to others. The word "location" is used in a literal sense—that is, to refer to the connection of person to place. In other words, people are identified according to their place of origin. Such an identification makes sense only in relation to other people from other places of origin. So people are identified as, for example, *orang Penyengat* (Penyengat people), *orang Pangkil* (Pangkil people), *orang Bintan* (Bintan people), and *orang Karas* (Karas people). These toponymic labels, however, have symbolic meaning only by reference to the hierarchy of ranks. For example, the term *orang Penyengat* derives its symbolic meaning from the rank with which it is associated. The same applies to the other toponymic labels.

The various toponymic labels we are discussing are hierarchically associated thus:

orang Penyengat ..... *raja, tuan said* (aristocrat)

orang Pangkil ..... *encik keturunan* (commoner)

orang Bintan ..... *keturunan Bintan* (commoner)

orang Karas[7] ..... *hamba raja* (serf)

The hierarchy of ranks is thus spatially realized. In a situation where there is no longer a sultanate—the last sultan having abdicated in 1911—such ranks have hardly any political substance. Yet they remain significant not as the feudal structure of an ongoing sultanate, but rather as a remembered order which exists as the-past-in-the-present, and which is spatially realized in the present. Since people are identified according to their place of origin, it follows that everyone classified as *Melayu* is located somewhere in this spatially realized past-in-the-present.

Because the *orang Penyengat* rank highest on the hierarchy, Penyengat itself is regarded as the symbolic center of *Melayu-ness* to which the other places are thereby peripheral. Penyengat is known as the "center of Melayu culture" (*pusat kebudayaan Melayu*). The relationship of Penyengat to the other *Melayu* places may be pictured as shown in Figure 11.3. This picture is very similar to that portrayed in Figure 11.2, with the vital difference that Tanjungpinang, the center of the material economy, is totally absent here, having been replaced by Penyengat, the symbolic center.

While it is theoretically possible to put forward a Cartesian assertion that the material economy and the symbolic economy are totally unconnected, it would, however, make more sense to argue the reverse. The same people are, after all, actively involved in both economies. On the assumption that they are not schizoid individuals, it would seem only commonsensical that they should make subjective connections between their material and symbolic involvement, such that an integrative habitus emerges.

Bourdieu (1977:77) discusses the nature of these subjective connections thus:

> Because the dispositions durably inculcated by objective condi-
> tions ... engender aspirations and practices objectively
> compatible with those objective requirements, the most
> improbable practices are excluded, either totally without
> examination, as unthinkable, or at the cost of the double nega-
> tion which inclines agents to make a virtue of necessity, that is,
> to refuse what is anyway refused and to love the inevitable.

Elsewhere, Bourdieu terms this process "euphemisation" (1977:191, 196) which is "the group's means of saving its 'spiritualistic point of honour'" (1977:196).[8]

In the context of Riau, a symbolic economy where the aristocratic *orang Penyengat* are located at the top of the social hierarchy may indeed be analyzed as an euphemization of a material economy where they are the other-dependent participants of a cash nexus over which they have no control. Other-dependence in the material economy is thus euphemized as self-sufficient independence in the symbolic economy. A scale of inverted values is thereby generated, such that those who are materially the most other-dependent are symbolically the most self-sufficient, and vice versa. According to this scale, the people of Pangkil and Bintan, who are located in the middle of the material continuum between the poles of other-dependence and self-sufficiency, are associated with the rank of "commoner" on the symbolic hierarchy. The people of Karas, who are located at the self-sufficient end of the material continuum, are associated with the rank of "serf" on the symbolic hierarchy.

As with material self-sufficiency, symbolic self-sufficiency derives from control over one's resources. But what constitutes symbolic resources? If the material continuum is marked by varying degrees of involvement in the cash nexus, what is the equivalent nexus in the symbolic economy? What is the symbolic currency? My informants speak of "purity" when they discuss the various ranks. The "aristocrats" are said to be the most *murni* (pure), the "commoners" less "pure," and the "serfs" are said to be *tidak murni* (impure). "Purity" is manifest in three areas of behavior: religion, language, and manners. Those who rank high are said to be "pure" because they are Islamic, because their language is refined, and because their manners are civilized. Those who rank lowly are said to be "impure" because they are un-Islamic, because their language is uncouth, and because their manners are uncivilized. The currency of the symbolic economy is thus a nexus of "purity" comprising Islam, *bahasa* (refined language), and *adat* (civilized manners). It is through the control of these symbolic resources that the "aristocrats" achieve their symbolic self-sufficiency.

As mentioned earlier, rank—particularly high rank—is transmitted patrilineally. Men are said to be "strong" (*kuat*) and women are said to be "weak" (*lemah*). Such a view dovetails with the domestication of women in the net-fishing communities. As I have suggested, this may be a means of euphemizing the material other-dependence of the men in such com-munities. Significantly, patrilineality is emphasized most in Penyengat, less so in Pangkil and Bintan, and least of all in Karas. Thus, even the kinship system is skewed by this inversion between the material and symbolic economies.

Just as the cash nexus does not involve all the people in Riau to an equal extent, so the symbolic nexus discussed earlier is more relevant to some and less to others. Not surprisingly, it is more relevant to those who benefit from it—namely, the people of Penyengat, Pangkil, and Bintan who may be adjudged "pure" according to this scale of values, and who may thereby symbolically disguise their material other-dependence. Such a scale of values is less relevant to the people of Karas who hold instead to a counter-image, an alternative symbolic economy, one in which the relevant currency is not "purity" but "indigeny." Interestingly, this alternative symbolic economy is also seen as derived from the past; however, the past that it harks back to is earlier than the reign of Sultan Sulaiman (1721–60).

In the context of our discussion, the significance of Sultan Sulaiman's reign is that he came to the throne only through the aid of some Bugis adventurers who were rewarded with the enfeoffment of political office. The position of *yang dipertuan muda* (he who is made junior lord), a position second only to that of the sultan himself, became a hereditary office corporately owned by the ennobled Bugis newcomers and subsequently by their descendants. Penyengat was the former capital of the *yang dipertuan muda*, and a large proportion of the Penyengat population are descendants of the enfeoffed Bugis nobles, who therefore bear the patrilineally transmitted title of *raja*. The rest of the Penyengat population are mostly descendants of court officials and the like. For my Penyengat informants, Sultan Sulaiman's reign is an important benchmark, because that was when their inherited rank was instituted.

In contrast to this view, however, my Karas informants look back beyond Sultan Sulaiman's reign to a time when the Bugis adventurers had not yet come, but a time when their own ancestors were already present in the Riau archipelago. They thus regard themselves as *asli* (indigenous) and hence prior to the Penyengat people who are regarded as *dagang* (foreign). The people of Pangkil who are associated with the rank of *encik keturunan* are also regarded as "foreign," the *encik keturunan* being the descendants of court officials serving the Bugis. The Bintan people are, however, accepted by my Karas informants as "indigenous," though regarded as no more than their equals. The alternative symbolic economy of the Karas people is thus not a hierarchy ranging from high to low, but rather a circle of belonging, with the "indigenes" located inside and the "foreigners" located outside.

As I have mentioned earlier, practically every able-bodied man on Karas island goes off on regular trading expeditions to places as far as Singapore and Jambi. Even the spear fishers are very wide-ranging in their travels, as they move from place to place, especially in accordance with the monsoon seasons.[9] In terms of their material economy they are very much outward orientated toward the wider world. Yet in terms of their symbolic economy, their orientation is inward, with such a value as "indigeny." Here again we have an inversion of values, whereby material outwardness is disguised by symbolic inwardness.

There are thus two symbolic economies—one based on a hierarchy of ranks inherited from a bygone sultanate, the other on a circle of belonging, enclosing some and excluding others. This symbolic duality would seem to be directly related to the material duality of other-dependence versus self-sufficiency.[10] The harmonization of the symbolic and the material economy generates what Bourdieu (1977) terms a "habitus," a habitual state of existence, a continuing set of dispositions. To the extent that different populations are involved in the two habitus discussed earlier, it would appear that these are what Bourdieu (1977:85) calls "class habitus"—that is,

> the system of dispositions (partially) common to all products of the same structures. Though it is impossible for *all* members of the same class (or even two of them) to have had the same experiences, in the same order, it is certain that each member of the same class is more likely than any member of another class to have been confronted with the situations most frequent for the members of that class.

Using the term "classes" implies that the different populations inhabiting the different habitus are nevertheless interacting within the same social reality. As pointed out earlier, the various communities we have discussed do not interact with one another in the context of the material economy. Their interaction is entirely symbolic. This may be explained thus.

The symbolic economy based on an inherited hierarchy of ranks pertains particularly to the people of Penyengat, Pangkil, and Bintan. Nevertheless, that symbolic hierarchy extends outward to include the people of Karas within its purview. The hierarchy and their attributed rank within it thus become a phenomenon the latter must deal with, insofar as it impinges upon their consciousness. This impingement occurs in two ways—through their own collective memory of their ancestors' rank in the era of the sultanate and through their own individual interactions with the people of Penyengat and Pangkil. (Bintan and Karas are located too far away from each other for much significant interaction to take place.)

In the first instance, it is hard for the Karas people to deny that their ancestors had the rank of "serf," since those ancestors are not so far removed in time from the present. In the second instance, the people of Penyengat and Pangkil are blatantly obvious about their scorn for the supposedly low-ranking Karas people. For example, the former would not even practice commensality with the latter, and would even go to the extent of bringing their own food whenever they visit Karas. The former also make fun of the way the Karas people talk, walk, and dress.

In these and other ways then, the people of Penyengat and Pangkil impose a symbolic domination over the Karas people. The alternative view that the Karas people have of themselves as "indigenous" and of the Penyengat and Pangkil people as "foreign" may thus be understood as the former's symbolic resistance to the latter.[11] What we thus have are symbolic classes and a symbolic class conflict.

Significantly, however, this symbolic class conflict is limited to the *Melayu* context from which the fish dealers, the sundry merchants, and the government bureaucrats are excluded. As mentioned, these people play dominant roles in the material economy of my informants, particularly at the other-dependent end of the material continuum. The absence of these materially dominant persons from the symbolic economy thus indicates a disjunction between symbolic classes and material classes.

From the perspective of those informants who are the most other-dependent, it is as if in lieu of giving overt recognition to their material submission to certain dominant others, that they exclude these materially dominant others from the symbolic economy and substitute instead a desire to dominate those who are not materially dominant but are materially more self-sufficient. In other words, a desire to dominate by symbolic means may be regarded as the misrecognition of an existing condition of material submission. *Melayu-ness* is thus defined symbolically precisely because it cannot be defined materially. In this light, ethnicity is a symbolic construction of identity that deliberately misrecognizes the material relations of domination-submission.

The Indonesian Context

Bourdieu (1977:183–197) argues that such symbolic misrecognition tends to occur in one particular mode of domination rather than in another:

> Once a system of mechanisms has been constituted capable of objectively ensuring the reproduction of the established order by its own motion . . . the dominant class have only *to let the system they dominate take its own course* in order to exercise their domination; but until such a system exists, they have to work directly, daily, personally, to produce and reproduce conditions of domination which are even then never entirely trustworthy. Because they cannot be satisfied with appropriating the profits of a social machine which has not yet developed the power of self-perpetuation, they are obliged to resort to *the elementary forms of domination*, in other words, the direct domination of one person by another. (Bourdieu 1977:190)

There are thus, on the one hand, a systemic mode of impersonalized domination, and on the other, an elementary mode of personalized domination.

It is questionable whether the Indonesian economy constitutes an impersonalized system of mechanisms, whereby profits can be appropriated without personal intervention. Geertz (1984:517) has summarized the debate over this question thus:

> Has or has not Indonesian history . . . consisted, from quite early on . . . of a progressive, step by irresistible step, encroachment of the logic of capitalism upon that of indigenous society such that society has been fairly thoroughly transformed into a commoditized, class polarized, "dependent" system, a peripheral outlier of a formerly colonial, now neo-colonial hierarchial world economy whose apex is, in Geoffrey Hainsworth's mocking phrase, "most likely located in the New York Board Room of the Chase Manhattan Bank"? Most . . . rather think that it has. Some . . . rather think that it has not. The difference of opinion is not, of course, whether such an impact has occurred and been extremely significant; no one, from any perspective, has ever denied that. It is whether the force of that impact has been such as to overwhelm Javanese rural society and "reconstitute" its peasantry in Capitalist, Man and Master terms, or whether it has been insufficiently massive or too specifically focused to overcome the "Asiatic" constraints proper to that society, the

immanent logic of the "Tributary" or the "Mercantile" or the "Feudal" Mode of Production. According to this way of thinking, the characteristic mark of capitalism is a fundamental opposition between the owners of the means of production and wage laborers, alienated from such ownership, while the characteristic mark of the Asiatic Mode of Production is one between patrimonial or feudal tribute-takers and the kin- and community-bound primary producers from whom the tribute is taken.

This question, however, cannot be answered without taking into account the political reality of the Indonesian nation-state. As several observers have noted, *bapakism*—literally, "fatherism"—is rife in Indonesian political life. The bureaucrat *bapak* (father) is the patron of a circle of citizen clients, his *anak buah* (children). These patron-client relationships are so pervasive that they constitute the very fabric of Indonesian politics. The nation as a whole is thus "composed of a multitude of groupings, each with its own elite and mass elements bound together by an elaborate set of personal relations, but only tenuously connected to surrounding social groupings" (Jackson 1978b:35).

As mentioned earlier, the Indonesian economy is subject to all sorts of arbitrary controls imposed by the bureaucrats (Booth and McCawley 1981:17–19.) Moreover, the basic form of government in Indonesia has been described as a "bureaucratic polity—that is, a political system in which power and participation in national decisions are limited almost entirely to the employees of the state, particularly the officer corps and the highest levels of the bureaucracy" (Jackson 1978a:3). Jackson (1978b:35) further notes that "only when a bureaucrat is a bapak can the official be sure that his orders will be carried out." If we consider these points in conjunction, then it seems quite clear that the Asiatic Mode of Production still dominates the Indonesian political economy, such that an "oriental despotism" manifests itself through a nationwide network of patrimonial appanages.[12] Following Bourdieu's argument cited earlier, it is precisely in such a situation that the elementary mode of personalized domination operates and symbolic misrecognition occurs.

I would thus suggest that the material dependence/symbolic independence of the Riau natives may be understood even in the larger Indonesian context. As mentioned, Riau Province is rich in oil, granite, bauxite, and tin. But these industries are not in the hands of the natives of the province. They are controlled by the *bapaks* of the Indonesian government and their favored companies. A dualistic material economy persists. As Hendra Esmara (1975) has pointed out, even though Riau is perhaps the richest province in Indonesia producing six-sevenths of Indonesia's crude oil, the oil revenue that accounts for

66 percent of the provincial income belongs to the central government and hardly reaches the natives of the province.

So even though the people of Penyengat, Pangkil, and Bintan may be involved in the cash nexus, their material economy is of a very small scale when compared to the national whole. In the larger Indonesian context, my informants in island Riau are located on the distant fringe of the national economy; they are materially very peripheral indeed. They are perhaps among the smallest *anak buah* of the smallest *bapak*. While this may be the case materially, my informants nevertheless do not see themselves as being symbolically peripheral to Jakarta. On the contrary, they see themselves as being symbolically central in *Melayu* Riau. In the larger national context, *Melayu* enthnicity may be regarded as the symbolic misrecognition of belonging to a materially peripheral Indonesian citizenry.

Acknowledgments

Field research in Riau was financed by a doctoral grant from the Australian National University. I am grateful for comments on this chapter by Sharon Siddique, Carl Trocki, my fellow authors in this volume, and other participants in the Michigan conference.

Notes

1. I must, however, add a caveat to the effect that the view of theory I adhere to is, to borrow the linguists' terms, "hocus-pocus" rather than "God's truth." In support of this view, Burling (1969:427) has stated:

   When a linguist makes his investigation and writes his grammar, is he discovering something about the language which is "out there" waiting to be described and recorded or is he simply formulating a set of rules which somehow work? Similarly, when an anthropologist undertakes a semantic analysis, is he discovering some "psychological reality" which speakers are presumed to have or is he simply working out a set of rules which somehow take account of the observed phenomena? . . . . It is always tempting to attribute something more important to one's work than a tinkering with a rough set of operational devices. It certainly sounds more exciting to say we are "discovering the cognitive system of the people" than to admit that we are just fiddling with a set of rules which allow us to use terms the way others do. Nevertheless, I think the latter is a realistic goal, while the former is not.

Although I am using one particular theoretical approach in this chapter, it does not imply that this is the only valid approach or that the same ethnographic data may not be analyzed in some other ways (see, for example, Wee 1985 for a somewhat different analysis of the topic).

2. For more ethnographic details, see Wee 1985.

3. The other districts are Pekanbaru (the provincial capital), Kampar, Indragiri Hulu, Indragiri Hilir, and Bengkalis. For more information on Riau Province as a whole, see Kato 1984.

4. If the word "headman-dom" seems awkward, that is only because it is a literal translation of the equally awkward Indonesian word *ke-penghulu-an*, which is a construction coined from *penghulu* (headman).

5. The following newspaper report appeared at the time of writing:

> The Indonesian government has ordered the reorganisation of the Customs Department in a drastic move to rid all ports and airports of corruption and bureaucratic delays. . . . More than 6,000 customs officials would begin indefinite paid leave tomorrow as part of the new sweeping measures. The announcement did not say how long the affected employees would stay away from work. Most imports and exports will now bypass Indonesian Customs and be cleared overseas by the independent Swiss-based surveyor, Societe General de Surveillance. Anchoring, towing, mooring and port storage charges will also be lowered and ships will no longer have to hire three government tugboats to dock. (*The Straits Times*, Wednesday, May 1, 1985)

It is, however, too soon to tell what effect these changes will have on the people of Riau.

6. These dates are derived from Raja Ali Haji (1982 trans.) and from Tengku Ahmad bin Tengku Abu Bakar 1972. Also see Andaya 1975 and Muchtar Lufti et al. 1977 for historical details.

7. Because the people of Nanga and Teluk Nipah are also associated with the rank of "serf" and because these two places are indeed located within the "headman-dom" of Karas, they will be subsumed under the name "Karas" in the rest of this discussion. However, the Penyengat people do tend to rank the people of Nanga and Teluk Nipah even lower than the people of Karas island.

8. See James Eder in this volume for another example of this process. The Philippine Negritos say they do not like farming in a context where the opportunities for efficient farming are being denied to them. They thereby refuse what is anyway refused.
9. During the northeast monsoon, the boat-dwelling spear fishers are mostly in the Lingga archipelago, and during the southwest monsoon, they are mostly in the Riau archipelago (see Map 11.1). This is not, however, a fixed pattern, for they are free to travel as far as they wish.
10. See Aram Yengoyan in this volume for a comparable case of dual hierarchies—one vertical and the other concentric. In a situation analogous to that in Riau, it is the more other-dependent Mandaya people who have a vertical structure of hierarchy, while the more self-sufficient Mandaya structure hierarchy in concentric circles.
11. See Alberto Gomes in this volume for another example of ethnic identity emerging as symbolic resistance to the domination of others. In Malaysia, Semai ethnicity seems to have arisen in opposition to Malay domination.
12. See, for example, Godelier 1977:99-124 on the Asiatic Mode of Production. For more details of the Indonesian political economy, see Mortimer 1973; Anderson and Kahin 1982.

References

Andaya, Leonard Y.
> 1975 *The Kingdom of Johor 1641–1728*. Kuala Lumpur: Oxford University Press.

Anderson, Benedict, and Audrey Kahin
> 1982 *Interpreting Indonesian Politics: Thirteen Contributions to the Debate*. Cornell Modern Indonesia Project, Interim Reports Series No. 62. Ithaca, NY: Cornell University.

Booth, Anne, and Peter McCawley
> 1981 The Indonesian economy since the mid-sixties. In *The Indonesian Economy During the Soeharto Era*, edited by Anne Booth and Peter McCawley. Kuala Lumpur: Oxford University Press. Pp. 1-22.

Bourdieu, Pierre
> 1977 *Outline of a Theory of Practice*. Translated by Richard Nice. Cambridge: Cambridge University Press.

Burling, Robbins
    1969 Cognition and componential analysis: God's truth or hocus-pocus?
        In *Cognitive Anthropology*, edited by Stephen A. Tyler. New York:
        Holt, Rinehart and Winston. Pp. 419–428.

Geertz, Clifford
    1984 Culture and social change: The Indonesian case. *Man* 19(4): 511–
        532.

Godelier, Maurice
    1977 *Perspectives in Marxist Anthropology*. Translated by Robert
        Brain. Cambridge: Cambridge University Press.

Hendra Esmara
    1975 An economic survey of Riau. *Bulletin of Indonesia Economic
        Studies* 11(3):25–49.

Jackson, Karl D.
    1978a Bureaucratic polity: A theoretical framework for the analysis of
        power and communications in Indonesia. In *Political Power and
        Communications: Indonesia*, edited by Karl D. Jackson and Lucian
        W. Pye. Berkeley: University of California Press. Pp. 3–22.
    1978b The political implications of structure and culture in Indonesia. In
        Political Power and Communications: Indonesia, edited by Karl
        D. Jackson and Lucian W. Pye. Berkeley: University of California
        Press. Pp. 23–42.

Kato, Tsuyoshi
    1984 Typology of cultural and ecological diversity in Riau. In *Transfor-
        mation of the Agricultural Landscape in Indonesia*, edited by
        Narifumi Maeda and Mattulada. Kyoto: Center for Southeast
        Asian Studies, Kyoto University. Pp. 3–60.

Mortimer, Rex (editor)
    1973 *Showcase State: The Illusion of Indonesia's "Accelerated Moder-
        nization."* Sydney: Angus and Robertson.

Muchtar Lufti, Suwardi MS, Anwar Syair, and Umar Amin (editors)
    1977 *Sejarah Riau*. Pekanbaru, Indonesia: Percetakan Riau.

Raja Ali Haji ibn Ahmad
    1982 *The Precious Gift: Tuhfat al-nafis*. An annotated translation by Virginia Matheson and Barbara Watson Andaya. Kuala Lumpur: Oxford University Press.

Sopher, David E.
    1977 *The Sea Nomads*. Singapore: Singapore National Museum.

The Straits Times
    1985 *The Straits Times*. Wednesday, May 1. Singapore: The Straits Times Press.

Tengku Ahmad bin Tengku Abu Bakar
    1972 *Sekelumit Kesan Peninggalan Sejarah Riau*. Dabo, Singkep. Typescript.

Wee, Vivienne
    1985 Melayu: Hierarchies of Being in Riau. Ph.D. dissertation, Australian National University, Canberra.

Wilkinson, Richard James
    1959 *A Malay-English Dictionary (Romanized)*. London: Macmillan. First published in 1932.

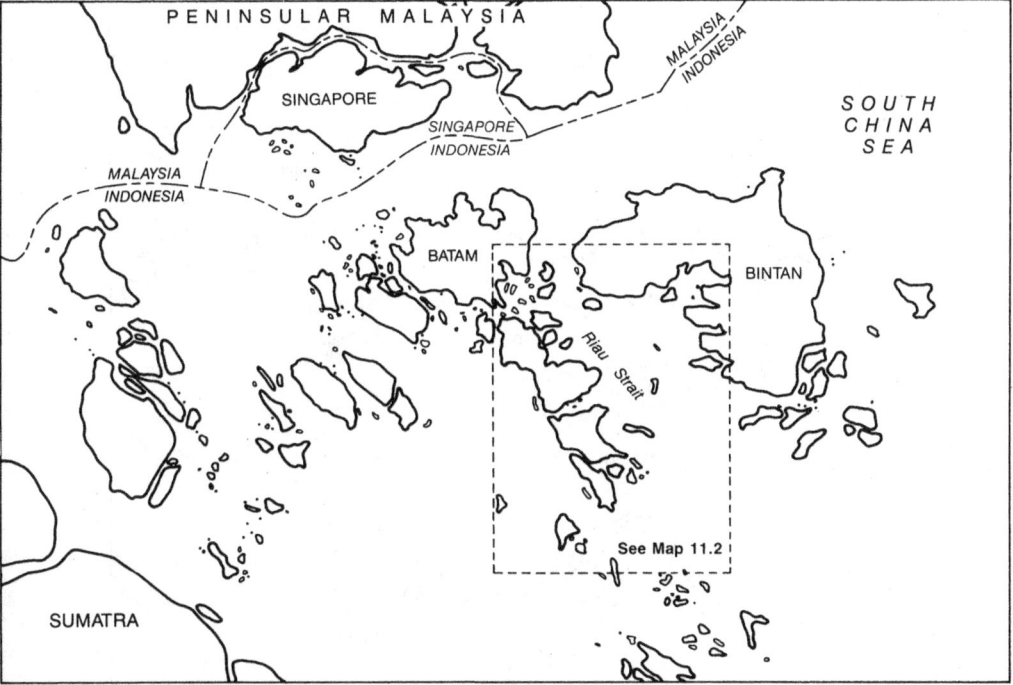

Map 11.1
Riau District

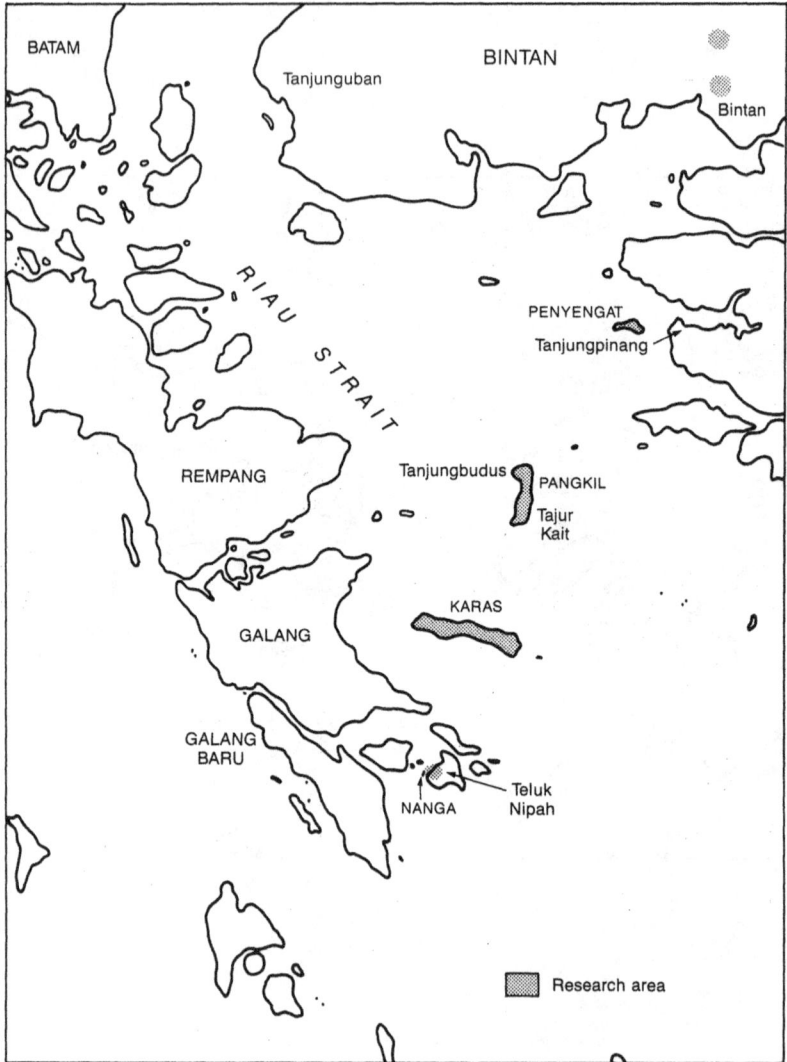

BATAM

BINTAN

Tanjunguban

Bintan

R I A U   S T R A I T

PENYENGAT
Tanjungpinang

REMPANG

Tanjungbudus   PANGKIL

Tajur
Kait

KARAS

GALANG

GALANG
BARU

Teluk
Nipah

NANGA

Research area

Map 11.2
Subdistrict of Southern Bintan

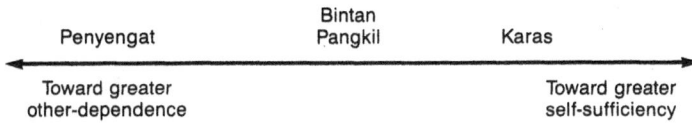

Figure 11.1
The Material Continuum

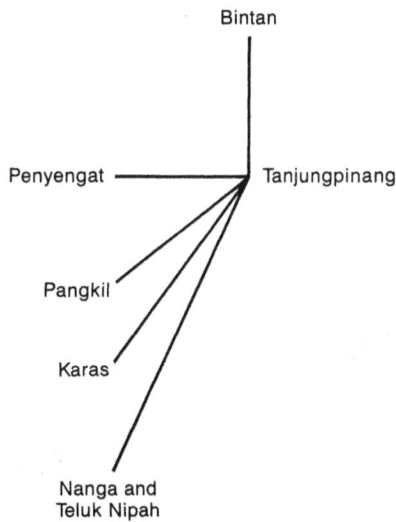

Figure 11.2
Tanjungpinang, the Center of the Material Economy

Bintan

Penyengat

Pangkil

Karas

Nanga and
Teluk Nipah

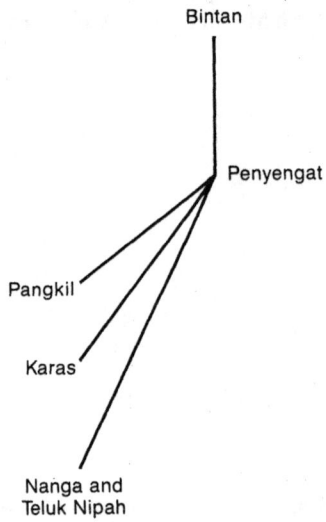

Figure 11.3
Penyengat, the Symbolic Center

APPENDIX

CONFERENCE PARTICIPANTS

Dr. Rowe V. Cadeliña
University Research Center
Silliman University
Dumaguete City, 6501
Philippines

Ms. Carol De Raedt
Cordillera Studies Center
University of the Philippines College Baguio
Baguio City 0203
Philippines

Dr. James F. Eder
Department of Anthropology
Arizona State University
Tempe, AZ 85281

Dr. Kirk Endicott
Department of Anthropology
Dartmouth College
Hanover, NH 03755

Dr. Brian L. Foster
Department of Anthropology
Arizona State University
Tempe, AZ 85281

Dr. Anan Ganjanapan
Department of Anthropology
Faculty of Social Science
Chiang Mai University
Chiang Mai, Thailand

Dr. Alberto G. Gomes
Department of Anthropology and Sociology
University of Malaya
Kuala Lumpur 22-11
Malaysia

Dr. Karl L. Hutterer
Museum of Anthropology
University of Michigan
Ann Arbor, MI 48109

Dr. Robert McKinley
Department of Anthropology
Michigan State University
East Lansing, MI 48824

Mr. Pei Sheng-ji
Yunnan Institute of Tropical Botany
Academia Sinica
Xishuangbanna
Yunnan Province
China

Dr. A. Terry Rambo
Environment and Policy Institute
East-West Center
1777 East-West Road
Honolulu, HI 96848

Dr. Renato Rosaldo
Department of Anthropology
Stanford University
Stanford, CA 94305

Dr. Vivienne Wee
Department of Sociology
National University of Singapore
Kent Ridge Campus
Singapore

Dr. Robert L. Winzeler
Department of Anthropology
University of Nevada
Reno, NV 89557

Dr. Aram A. Yengoyan
Department of Anthropology
University of Michigan
Ann Arbor, MI 48109